"Better than Heloise's Hints for a Healthy Planet."
—*Library Journal*

"Destined to become a classic."
—Debra Lynn Dadd, author, *Nontoxic & Natural*

"The ultimate in childproofing. Every home needs this veritable encyclopedia of nontoxic, environmentally friendly cleaning products. . . ."
—Wendy Gordon Rockefeller,
co-chair, Mothers and Others For a Liveable Planet

"A boon for environmentally sensitive patients."
—Ken Bock, M.D., environmental and preventative medicine

"Clean & Green is the 'Joy of Cleaning'."
—Nikki Goldbeck, nutritionist/author

"I never thought a book (or anything else) could make me enthusiastic about housecleaning—but this book has!"
—Donella Meadows, nationally syndicated columnist and
adjunct professor of environmental studies, Dartmouth College

"Annie Berthold-Bond's book provides the means to protect our environment and our health and to say 'no' to the overuse of household chemicals."
—Irene Caldwell, Household Hazardous Waste Coordinator

CLEAN & GREEN

The Complete Guide

To Nontoxic And Environmentally Safe

Housekeeping

Annie Berthold-Bond

Foreword by Debra Lynn Dadd

Ceres Press
Woodstock, New York

Published by:Ceres Press
            PO Box 87
            Woodstock, New York 12498

Cover painting: Rosemary Fox

Cover design: David Goldbeck and Peter Michos

Art Director: Alan McKnight

Printing: 12
Updated Edition

Library of Congress Catalog Card Number: 90-082580

Berthod-Bond, Annie
ISBN: 1-886101-01-9

DISCLAIMER
   Neither Ceres Press nor Annie Berthod-Bond is responsible for any damages, loses, or injuries incurred as a result of any of the material contained in this book.
   The author has attempted to include cautions and guidelines for using ingredients in this book. It is impossible however, to anticipate every conceivable use of formulas or their reactions on different materials or on different people. For this reason, we cannot assume responsibility for the effects of using the formulas or other information found here. It is recommended that before you begin to use any formula or substance you read the directions carefully. Always test a formula first in an inconspicuous place. If you have any questions conerning the safety or health effects of any formula, consult first with a physician or other appropriate professional.

PRINTING NOTE
Printed on acid-free recycled paper with vegetable based ink.
Printed in Canada.

Foreword

Every once in a while I come across a book that I wish I had written. *Clean & Green* is one of these. I found out about it as I was writing the cleaning products chapter of my book *The Earthwise Consumer Guide*, and I wanted to just take *Clean & Green* and put it, in its entirety, right in the space in my book where my cleaning products chapter goes.

I've never seen such an impressive array of formulas for making your own cleaning solutions from natural, healthful, and environmentally-safe substances. A lot of creativity went into this book. Instead of saying "Here's a toxic cleaning product that works - how can we make it safer?" Annie has said, "OK, these substances are safe - how can we use them for cleaning?"

As we all make changes in the way we do things for the good of our Earth, this book will be an excellent reference to help us reduce the manufacture, use, and disposal of toxic substances - especially when we need to find a way to clean something where the solution isn't obvious. Before this, who knew how to make their own car wax? *Clean & Green* is destined to become a classic. It belongs in every household.

- *Debra Lynn Dadd*

For Kathy and Dave

Acknowledgements

I'd like to give a special thanks to my parents, who by example, taught me to respect the earth.

I'd also like to thank Phillip Dickey, of the Washington Toxics Coalition, for his thoughtful reading of the manuscript, and Kathy Gibbons, Ph.D., for her technical assistance relating to chemistry.

A heart-felt thanks to Nikki and David Goldbeck. Their perceptions have been astute and their editing has been invaluable. Thanks are due also to Nancy Baker for her professional and attentive work on the manuscript.

Without my husband Daniel, this book would never have been written. On a practical level he has put in double overtime to help me have the time to work. On an emotional level he has put in triple overtime to help me try to become all the things I can be.

As to my daughter Lily, without her I may never have developed the determination to write the book. In the end, it is for her and her grandchildren, and for all of her friends and cousins and their grandchildren's sake, that I wrote Clean & Green.

CONTENTS

INTRODUCTION

I began doing research into practical nontoxic care of the home because I was sick from being exposed to pesticides. This required that I be in an environment that was free of synthetic chemicals. If the furniture needed polishing, I couldn't use a commercial furniture polish, but had to find safe alternatives that worked. Gradually, over a number of years, I began compiling workable solutions to cleaning problems. I found a few acceptable commercial products, but most were homemade formulas that, through experience, I found met my needs.

In the beginning, my emphasis regarding nontoxic living was on personal health, but as time went on I became more aware, like many of us, that the ecosystems of the earth were threatened. Something I'll never forget is the day of a massive herbicide spraying done to kill dandelions on the lawns of a campus where I lived. The campus is located on one of the richest migratory paths for birds in the world, and the evenings were so loud with bird songs that it was hard to hear oneself think! On the evening of the spraying the campus was bound in, as if in a heavy fog, with one of the eeriest silences I have ever heard. I didn't hear a bird sing for three days. Rachel Carson, author of *Silent Spring*, was right: The birds really do leave when they are exposed to toxic chemicals. That night of the herbicide spraying, nontoxic living became more than just a matter of protecting my health. I understood in a very poignant way what we were doing to the earth.

It is now a number of years after that silent night, and I find that people are becoming increasingly vocal about wanting both to protect their health and heal and rejuvenate our planet. And for good reason. According to an article in the *Christian Science Monitor* (January 23, 1990), "A surprising number of products used in the house and garage are toxic..." and there's no question that some of these products end up in our water and air, although no one knows exactly how much. The Clean Water Fund, a non-profit organization, estimates that the average American uses forty pounds of unsafe household cleaners each year. Multiply that number by 245 million Americans (and other users worldwide) and the effect is significant. If that's not bad enough, studies have shown that the greatest number of poisonings are from cleaning substances and 64 percent of the victims are under the age of six.

The consequences become even greater if you consider that toxic products are also produced in the manufacturer of these cleaning substances. Spot removers, furniture polishes, toilet, drain, and oven cleaners are products we use every day that have both a significant health and environmental impact.

A common problem seems to be that people don't know where to start, what to do, or how to use nontoxic products. A person might believe that using nontoxic products to protect our landfills from household hazardous waste is important, but that same well-intentioned person will say, "I feel badly that I bought an oven cleaner, but what do I do when I'm having a dinner party and the oven is smoking from burnt grease and I have to clean it in a hurry?" People don't know what the accessible alternatives are. Having been at a loss for how to clean the oven with nontoxic products at one time myself, I can help out. With this handbook I have provided you with practical, hands-on, day-to-day help in nontoxic and environmentally safe household cleaning.

I have discovered that cleaning with natural materials takes on a simplicity similar to that of the three canisters many people have on their kitchen counters containing flour, sugar, and tea. One might instead have canisters of baking soda, washing soda, and borax, which can become as integral a part of one's life as the food canisters. Just like in making a cake, if you use the right ingredients at the right time, you have a successful result. Further, the more I discovered about the remarkable cleaning powers of minerals, the more I learned about the planet and its geology, geography, and chemistry. In learning about cleaning with vegetable-oil-based soaps, I started to think of gardens and growing things. I started to feel interconnected with the planet, and I have become increasingly educated about "the way of things". In contrast, commercial cleaning products conjure up nothing but images of synthetics, unpleasant smells, rashes, headaches, shopping malls, parking problems, and offensive advertising.

Many people who have become sick from being exposed to chemicals have been told about the barrel analogy, and it is worth repeating because it helps make the risks to health from chemicals clear. Imagine yourself to be an empty barrel. Each pollutant that you breathe, eat, or absorb through your skin fills a part of the barrel. If you paint your house, your barrel will no longer be empty. You might wax your floors, spray your rose bushes, and do all the other things we innocently do. Each one of these pollutants adds a bit more to the barrel. How fast the barrel fills is different for each of us, depending on our biochemical makeup, the chemical ingredients that are filling the barrel, and what our past exposures have been. When the barrel overflows we get sick. The furniture polish, the leaky oil burner, the dishwashing liquid - these are all pollutants that our bodies must somehow clean from our systems. These are pollutants that add up. If we reduce the number of environmental contaminants brought into the home, we do a lot toward preventing a "full barrel." If our house or apartment is free of toxins then the body rests, heals, and recuperates. We can start *emptying* the barrel.

As to the environment, one of the bleakest trips that you can take is to ride the train from New York City to Washington, D.C. The train winds through the murkiest, muckiest, most discolored earth you could imagine. The tracks are lined with refineries and smokestacks. Being on the train, the one thing that you miss is the smell of the air. The manufacturers causing this devastation are producing products that we use at home: paints, furniture polish, laundry soap. What we throw away of these products goes into our landfills, and from there it can leach into our water. The same goes for what we wash down the drain.

With a small amount of adjustment by all of us, our lakes and streams can run clear again. If there is enough of a grass-roots outcry, then manufacturers will start producing truly environmentally safe products. Pesticide companies will develop safe pest control. Scientists will search for solutions. If we reject products that become hazardous waste and pollute our waters, manufacturers will have no reason to produce them. If we start now it is still possible to protect our planet.

A memory of my childhood is the spring smell of melting snow, clean air, and the earth. The air is sweet and cool, and crocuses start pushing up through the leaves. We play marbles in the driveway and have rosy cheeks. In the present, my daughter watches in wonder as the song birds come to the bird feeder at the window. Do I tell her that song birds are becoming extinct? About global warming and hot summers? That fresh air may be a thing of the past? When I was scared of nuclear war and couldn't sleep, my father always reassured me with the thought that no one was crazy enough to push the button. My hope is that with a commitment to protect the earth on the part of all of us, we will bring real change, and my daughter will be able to reassure *her* children that the earth is our home and we aren't crazy enough to destroy it.

Annie Berthold-Bond
Annandale, New York
May, 1990

Chapter 1

ECOLOGICAL CLEANING

Nontoxic Cleaning: Does It Work?

Yes, yes, and again, yes! I was not easily convinced on this issue because I believed that chemical companies had secret formulas that couldn't be duplicated and I thought their products must be superlative just because they were sold in stores. On researching nontoxic cleaning and reading hundreds of folk recipes, I started to see that it was the folk recipes that had the secret ingredients and that chemical companies were just copying the folk recipes, using synthetic chemicals instead. The acid in lemon juice works as well as an artificial acid to clean aluminum, and a lemon doesn't threaten your health. The "secret" of this handbook is that in many cases the recipes are the old-fashioned version of modern synthetic formulas.

Time and Finances

I need to dispel a few myths about the time and cost involved in nontoxic cleaning. People have a sense that it is expensive, time consuming, and inconvenient.

Using the recipes in this book will cost you pennies in comparison to the many dollars spent on commercial products. Some recipe ingredients are free: You can make a batch of metal cleaner when tomatoes are ripe in your garden, or when it's time to clean the aluminum you can use the discarded lemon peel after squeezing the juice for salad dressing. For a nominal amount of money, I bought the ingredients to make furniture, car and floor wax, and this one batch of supplies will make enough wax for many years to come. There is not one recipe I recommend that is even remotely as expensive as its commercial counterpart. Recycled materials are suggested in this handbook, too. A colorful piece of cloth recycled into a rag is softer and a lot more pleasant to touch than store-bought "window wipes" - and a rag is free. I actually have a special cleaning rag. It's my favorite color, one hundred percent cotton, and worn to a flannel softness. Washing it and hanging it to dry is a pleasure. It's not possible to feel that way about a paper towel.

At the end of this handbook I have a chapter that lists the names of manufacturers of environmentally responsible consumer products. Shopping for them is easy as they are readily available in health food stores and from mail order suppliers. These products are not necessarily more costly than supermarket brands but are more expensive than homemade formulas. The laundry soap I use is a vegetable-based, biodegradable, safe soap made from renewable resources. The manufacturer recommends using only one ounce per load of laundry, so one bottle lasts a long time. Having switched to this soap from a popular supermarket brand, I am actually saving money. I recommend talking to the salespeople as they can often help you find the best buy. I also highly recommend buying products in bulk. You can save even more money, save a lot of time shopping, and cut way back on unnecessary packaging.

There are also some useful accessories you can buy that give nontoxic cleaning simplicity and convenience. For example, since spray bottles are so expedient to use, I suggest buying a few and putting your nontoxic formulas in them. If you like spraying a cleaner onto your bathtub tiles and wiping it up, there is no reason why you have to change because you are now using nontoxic ingredients. There is more on containers and utensils later in this chapter.

Once you have the supplies in your cupboard, cleaning without synthetic chemicals is easy. You are just grabbing a different bottle than before. It takes no more time. Readjustment may be needed in obtaining and understanding your new supplies, but that is all. Once this initial step is taken, in a very short time you will find nontoxic cleaning *more* convenient. You will start integrating nontoxic agents and also begin to learn that if you are out of vinegar for a formula you can use a lemon, or a grapefruit, or some rhubarb.

Commercial Products: What to Eliminate and Why

There has been very little conclusive research into how many commercial products affect health and even less on how they affect the environment. For this reason alone, I recommend approaching commercial products with caution. The fact that a product is on the shelf of your grocery store is not proof of acceptability. For years, industry has been selling products before adequate tests were available to determine whether the ingredients used were carcinogens. Environmental impact studies are, at best, rare. Some people estimate that only three percent of carcinogens are known and registered - and cancer is only the tip of the iceberg when it comes to health effects of chronic low-level exposure to neurotoxic poisons. When using the following table, be aware that it only documents what is *known*.

COMMERCIAL PRODUCTS: COMMONLY USED INGREDIENTS, HEALTH EFFECTS, AND ENVIRONMENTAL IMPACT

COMMERCIAL ALL PURPOSE CLEANERS, ABRASIVE AND LIQUID

May Contain: Complex phosphates, chlorinated phosphates, dry bleach, kerosene, morpholine, petroleum-based surfactants, sodium bromide, glycol ether, Stoddard solvent, EDTA, naphtha.

Toxicity: Chlorinated materials can form organo-chlorine compounds, which in turn are stored in fat cells and can enter mothers milk. Morpholine is very toxic and a liver and kidney poison. Glycol ether, Stoddard solvent, naphtha, and kerosene are neurotoxins and central nervous system depressants which can cause confusion, headaches, lack of concentration, and symptoms of mental illness. Glycol ether is also a kidney and liver poison. Sodium bromide can cause confusion.

Environmental Impact: Stoddard solvent, petroleum distillates (naphtha), and very toxic materials such as morpholine are considered hazardous waste and cause pollution. Phosphates cause algae bloom, and chlorinated materials can form other compounds (like DDT) which break down slowly in the ecosystem as pollutants and are stored in the fatty tissue of wildlife. EDTA binds with heavy metals in our lakes and streams and thereby activates the metals. Petroleum-based products are nonrenewable resources, are slow to break down in the environment, and can contain impurities which contaminate air and water.

COMMERCIAL DEODORIZERS

May Contain: Methoxychlor, petroleum distillates, formaldehyde, p-Dichlorobenzene, piperonal butoxide, o-phenylphenol, aromatic hydrocarbons, salicylates, naphthalene.

Toxicity: Methoxychlor is a chlorinated hydrocarbon pesticide which accumulates in fat cells and can enter mothers milk. It is also neurotoxic and can over stimulate the central nervous system. P-Dichlorobenzene, aromatic hydrocarbons, and naphthalene are central nervous system depressants and neurotoxins which can cause headaches, lack of concentration, confusion, and symptoms of mental illness. Salicylates can cause strong allergic reactions and are very toxic. Formaldehyde is assumed to be carcinogenic. Piperonal butoxide is a suspected carcinogen.

Environmental Impact: Chlorinated hydrocarbons (like DDT) break down slowly in the environment. Petroleum distillates, aromatic hydrocarbons, formaldehyde, and very toxic materials such as salicylates are considered hazardous waste. Aromatic hydrocarbons contain impurities which contaminate air and water. Petroleum-based products are nonrenewable resources, are slow to break down, and can contain impurities which contaminate air and water.

COMMERCIAL DISH (HANDWASHING) DETERGENT

May Contain: Petroleum-based surfactants, naphtha, germicides, chloro-o-phenylphenol, diethanolamine, complex phosphates, sodium nitrites.

Toxicity: Chloro-o-phenylphenol is very toxic and a metabolic stimulant. Diethanolamine is caustic and a possible liver poison. Naphtha is a central nervous system depressant and a neurotoxin which can cause headaches, lack of concentration, confusion, and symptoms of mental illness.

Environmental Impact: Complex phosphates cause algae bloom. Naphtha and very toxic materials such as chloro-o-phenylphenol are considered hazardous waste. Chlorinated materials can form organo-chlorine compounds which are pollutants that break down slowly in the ecosystem, and are stored in the fatty tissue of wildlife. Petroleum-based products are nonrenewable resources, are slow to break down in the environment, and can contain impurities which contaminate air and water.

COMMERCIAL DISINFECTANTS

May Contain: Naphtha, butyl cellusolve, chlorinated germicides, sodium hypochlorite, sodium sulfite or nitrite, petroleum-based surfactants.

Toxicity: Aphtha is a neurotoxin and central nervous system depressant which can cause confusion, headaches, lack of concentration, and symptoms of mental illness. Butyl cellusolve is highly toxic and sodium nitrite is extremely toxic. Sodium sulfites can cause death in asthmatics.

Environmental Impact: Naphtha sodium hypochlorite, butyl cellusolve, and highly toxic materials are considered hazardous waste. Sodium hypochlorite and chlorinated germicides can also form organo-chlorine compounds (like DDT) which are pollutants that break down slowly in the ecosystem and are stored in the fatty tissue of wildlife. Petroleum-based products are nonrenewable resources, are slow to break down in the environment, and can contain impurities which contaminate air and water.

COMMERCIAL FURNITURE POLISHES

May Contain: Napthas and other petroleum-distillates, propellants, diglycol laurate, amyl acetate, petroleum based waxes, mineral spirits.

Toxicity: Naphtha, diglycol laurate, amyl acetate, and mineral spirits are neurotoxins and central nervous system depressants which can cause headaches, confusion, lack of concentration, and symptoms of mental illness. Diglycol laurate can also be a liver and kidney poison. Mineral spirits can cause upper respiratory tract irritation and can contain impurities, including the carcinogen benzene.

Environmental Impact: Napthas and mineral spirits are considered hazardous waste. Petroleum-based products such as mineral spirits are nonrenewable resources, are slow to break down in the environment, and can contain impurities which contaminate air and water.

COMMERCIAL GLASS CLEANERS

May Contain: Organic solvents, petroleum-based waxes, complex phosphates, ammonia, phosphoric acid, alkyl phenoxy polyethoxy ethanols, naphtha, butyl cellosolve.

Toxicity: Organic solvents, naphtha, and petroleum-based waxes are neurotoxic and central nervous system depressants which cause headaches, lack of concentration, confusion, and symptoms of mental illness. Organic solvents also commonly contain impurities such as the carcinogen benzene and can cause respiratory irritation. Butyl cellusolve is very toxic. Phosphoric acid and ammonia are highly corrosive.

Environmental Impact: Organic solvents, naphtha, petroleum-based waxes, ammonia, and butyl cellusolve are considered hazardous waste. Complex phosphates cause algae bloom. Petroleum-based products are nonrenewable resources, are slow to break down in the environment, and can contain impurities which contaminate air and water.

COMMERCIAL LAUNDRY DETERGENT

May Contain: Petroleum-based sufactants of the aryl and alkyl group, tetrapotassium pyrophosphate, complex phosphates, fluosilicate, sodium toluene and xylene sulfonate, EDTA, optical brighteners, benzethonium chloride.

Toxicity: Tetrapotassium pyrophosphate is highly toxic, corrosive and irritating, and is suspected of forming organophosphate properties. Fluosilicate is so toxic it is used as a pesticide. Benzethonium chloride is also highly toxic. Optical brighteners can cause strong allergic reactions when exposed to sunlight.

Environmental Impact: Complex phosphates cause algae bloom. EDTA binds with heavy metals in our lakes and streams and thereby activate the toxic metals. Optical brighteners can cause mutations in bacteria and do not biodegrade well. Chlorinated materials can form organo-chlorine compounds which are pollutants that break down slowly in the ecosystem and are stored in the fatty tissue of wildlife. Petroleum-based products are nonrenewable resources, are slow to break down in the environment, and can contain impurities which contaminate air and water.

COMMERCIAL METAL POLISHES

May Contain: Perchloroethylene, chromic acid, plasticizers, silver nitrate, phenolic derivative, kerosene, synthetic waxes, chromic acid, naphtha, other organic solvents.

Toxicity: Perchloroethylene, kerosene, naphtha, chromic acid, and organic solvents are neurotoxic and central nervous system depressants which can cause headaches, confusion, lack of concentration, and symptoms of mental illness. Perchloroethylene is an assumed carcinogen, a liver and kidney poison, and can cause death. Further, perchloroethylene is a chlorinated hydrocarbon and is stored in fatty tissue and from there can enter mothers milk. Silver nitrate is highly toxic and corrosive. Chromic acid is extremely toxic, a liver and kidney poison, and a possible carcinogen.

Environmental Impact: Most of the ingredients of metal polishes are considered hazardous waste. Chlorinated materials can form organo-chlorine compounds which are pollutants that break down slowly in the ecosystem and are stored in the fatty tissue of wildlife. Petroleum-based products are nonrenewable resources, are slow to break down in the environment, and can contain impurities which contaminate air and water.

COMMERCIAL OVEN CLEANERS

May Contain: Ether-type solvents, petroleum distillates, methylene chloride, butyl cellusolve, lye.

Toxicity: All of the above ingredients except lye are neurotoxic and central nervous system depressants which can cause headaches, confusion, lack of concentration, and symptoms of mental illness. Methylene chloride is a chlorinated hydrocarbon which is stored in fatty tissue. Further, methylene chloride is a liver and kidney poison. Lye is a corrosive poison. Ether-type solvents can contain impurities including the carcinogen benzene, and can cause respiratory distress.

Environmental Impact: Ether-type solvents, petroleum distillates, methylene chloride, butyl cellusolve, and lye are considered hazardous waste. Chlorinated materials can form organo-chlorine compounds which are pollutants that break down slowly in the ecosystem and are stored in the fatty tissue of wildlife. Petroleum-based products are nonrenewable resources, are slow to break down in the environment, and can contain impurities which contaminate air and water.

COMMERCIAL SPOT REMOVERS

May Contain: P-hydroxybenzoic acid, oxalic acid, naphtha, benzene, perchloroethylene or trichloroethylene, sodium hypochlorite, hydrofluoric acid, aromatic petroleum solvents, aliphatic hydrocarbons, chlorinated hydrocarbons, other petroleum hydrocarbons.

Toxicity: The ingredients listed above are so threatening to health that spot removers should simply not be used. Many of the above ingredients are carcinogenic, suspected of being carcinogenic, neurotoxic, central nervous system depressants, stored in fatty tissue, a cause of respiratory distress, liver and kidney poisons, extremely toxic and corrosive, and can cause death.

Environmental Impact: All the ingredients listed are considered hazardous waste. Chlorinated materials can form other compounds which break down slowly in the ecosystem and become pollutants and are stored in the fatty tissue of wildlife. Petroleum-based products (especially ones as toxic as these) are nonrenewable resources, are slow to break down in the environment, and can contain impurities which contaminate air and water.

COMMERCIAL TOILET BOWL CLEANERS

May Contain: Complex phosphates, o-or-p Dichlorobenzene, chlorinated phenols, kerosene, salicylates, germicides, fungicides, 1, 3- Diochloro-5, sodium acid oxalate, sodium acid sulfate.

Toxicity: Sodium acid oxalate, chlorinated phenols, and o-or-p-Dichlorobenzene are highly toxic. Sodium acid sulfate is highly corrosive. Chlorinated phenols are not only corrosive but metabolic stimulants. Fungicides can cause liver and kidney damage. O-or-p-Dichlorobenzene is a liver and kidney poison, as well as being a powerful central nervous system depressant which can cause headaches, lack of concentration, confusion, and symptoms of mental illness. Germicides can be toxic.

Environmental Impact: Highly toxic materials like sodium acid oxalate, chlorinated phenols, and o-or-p-Dichlorobenze are considered hazardous waste. Chlorinated materials can form other compounds which break down slowly in the ecosystem and become pollutants, and are stored in the fatty tissue of wildlife. Petroleum-based products are nonrenewable resources, are slow to break down in the environment, and can contain impurities which contaminate air and water.

Resources:
Robert E. Gosselin, M.D., Ph.d., Roger Smith, Ph.D., and Harold C. Hodge, Ph.D., D.Sc., *Clinical Toxicology of Commercial Products*, 5th Edition. Baltimore: Williams & Wilkins, 1984.
The Ulster County Environmental Management Council Household Hazardous Waste Task Force, *Report to the Ulster County Legislature Community and Environmental Affairs Committee*. Kingston, New York: Environmental Management Council, 1989.
The U.S. Department of Health and Human Services, *Fifth Annual Report on Carcinogens Summary 1989*. Washington: GPO, 1989.
The U.S. Office of Technology Assessment, *Neurotoxicity: Identifying and Controlling Poisons of the Nervous System Summary*. Washington: GPO, 1990.

Commercial Products: How to Eliminate Them

1. I suggest you make a pile of all the unacceptable cleaning agents in your home and garage. While you are at it, include cans of old pesticides and paints, too. For many products, I don't recommend "using everything up" instead of disposal, as many recycling centers often do, because there is increasing evidence that many commercial products carry health risks. Pesticides, herbicides, solvents, and oven cleaners are examples. Moreover, a person may have a can of something in their garage that, without their being aware, has long since been taken off the market for safety reasons. By using it up they may incur injury. However, I am not advocating people take items like laundry detergents to household hazardous waste centers. I suggest using the preceding chart as a guide to determine which products are totally unacceptable to have in the house or landfill and getting at least these to a household hazardous waste center.

2. Call your county health department to find out how to store household hazardous waste temporarily and to find out when there will be hazardous waste pickups.

A few tips:

Never mix different products together, as new and deadly chemicals can be made inadvertently.

Store household hazardous waste away from animals, children, and heat. Antifreeze has a sweet taste, for example, and animals like to eat it.

Don't pour any of this material down the drain or throw it into your regular trash. It not only may be hazardous waste but may be very dangerous to your health.

If you spill anything, try to absorb it with unperfumed, pure clay kitty litter or, yes, a disposable diaper, as these two items don't react with chemicals. Keep children and pets away from the area and try to ensure as much ventilation as possible. These directions apply only to household products. For a spill of industrial-strength chemicals, call the police.

3. Be very cautious of containers, too, especially the ones that contained pesticides.

The Transition

In order to make the transition to nontoxic living, the first thing I would suggest you do is to take a deep breath and decide that your concern about your health and the environment is strong enough to change your habits. Once you have done that, cleaning in a different way won't be hard. I find nontoxic cleaning rewarding - and cleaning the house has never been high on my list of things I like to do. The idea of cleaning the bathroom with baking soda and freshly picked rose petals is so pleasing that as unbelievable as it sounds, sometimes I actually *want* to go and do it! The furniture polish I make evokes smells and images I have conjured up when reading 19th-century novels. The wood has a clean, rich, nutty smell. Sleeping on detergent-free sheets seems more restful. Not absorbing chemical residue through my skin when I wash the dishes is reassuring and its good to know there are virtually no bottled poisons that children could drink and that the air isn't full of offensive chemicals. My home has virtually become a nontoxic sanctuary.

Minerals, Plants and Animal: Natural Ingredient Choices

In marked contrast to the commercial products listed in the preceding table, nontoxic cleaning will introduce you to an abundant array of the earth's resources. Ingredients used in this book's recipes are natural products of our planet: renewable, nonpolluting, harmless to our groundwater, and safe for our health, our fellow creature's health, and future generations. The recommended commercial products, too, are natural, nontoxic, and environmentally safe.

An Introduction to Natural Ingredients

The key to successfully making your own cleaners is to understand the way the ingredients work. Equipped with this knowledge you can approach cleaning problems with the comforting feeling that you are using the right thing at the right time. The range and breadth of natural cleaning materials is impressive. The following is a list of some commonly used materials, their unique properties, and recommendations for use as cleaners. For more specific cleaning information, consult the appropriate chapters in this book.

Minerals

Alum: Cleans stains such as rust on porcelain and lifts away hard water spots. Particularly effective when mixed with vinegar or lemon juice.

Baking Soda: Odor absorbing, deodorizing, mild abrasive.

Borax: Disinfects, deodorizes, inhibits mold growth.

Chalk: A non-abrasive cleaner often used as a whitener.

Cream of Tartar: Cleans porcelain, drains, and metal.

Pumice: Effective against stains, increases the sudsing quality of soaps, moisture absorber. Note: The fine powder should not be inhaled.

Salt: An effective non-scratching abrasive cleaner with bacteria inhibiting qualities.

Sodium Percarbonate: A natural bleach for white items.

Washing Soda: Cuts grease, cleans petroleum oils, cleans dirt. Washing Soda is slightly caustic. I recommend wearing gloves when using this ingredient.

Zeolite: I am pleased to introduce zeolite to the list of safe cleaning supplies. Zeolite is a one hundred percent nontoxic, naturally occurring mineral found near volcanic activity. What makes zeolite unusual is that it is the only mineral that is an ion exchanger in its natural state, which means it naturally absorbs pollutants from the air. It can absorb ammonia from water and clear an entire house of severe smoke damage in three days. To do this without the use of toxic chemicals or ozone generators is almost unheard of. It is claimed that zeolite will also extend the shelf life of fruit and vegetables for as long as three weeks. G&W Supply (which has the purest form of zeolite I know of) offers it for sale in granular form and in three sizes of "breather bags" that you can place in closets or other problematic areas. The zeolite itself is available in four dimensions from fine talc (which is dusty) to larger pellets ⅜ inch in diameter. Zeolite also "de-sorbs." Just put it in the sun, the absorbed fumes will dissipate, and you can use it again and again. Zeolite is also a water softener.

Plant Materials

Antiseptic Herbs: Some herbs have bacteria inhibiting properties. They include heather, myrrh, peppermint, eucalyptus, lavender, thyme, mint, and clove.

Australian Tea Tree Oil: This essential oil is a broad spectrum fungicide and bactericide.

Carnauba Wax: The hardest natural wax known to man comes from a Brazilian palm tree.

Citrus (lemons, limes, oranges, grapefruit): These citrus fruits have stunning cleaning properties. Grapefruit seed extract is considered antifungal, antimicrobial, and antibacterial. It is used as a preservative in cosmetics. Citrus-peel extracts are natural solvents, flea deterrents, and cleaners. Some of the most sophisticated research being done in the area of nontoxic flea control is on components in citrus peel called linalool and d-limonen. The latter, particularly, can kill all stages of the flea from egg to adult. For natural cleaning recipes, consider lemon juice as an acid-type cleaner for mineral buildup, tarnish, and grease. Lemon juice is also a remarkable air freshener; add some to your cleaner. To find out how to make lemon oil, see Essential Oils, below.

Essential Oils: Essential oils are the essence of a plant's fragrance and are used as the bases of perfumes. The fragrance is extracted from the plant by steam distillation, which produces the purest essence, or by soaking the plant matter in oil or alcohol. Note that these oils are so strong they can be dangerous and must be used sparingly. Be cautious of store-bought essential oils as many of them contain synthetics. Karen's Nontoxic Products is known to carry some pure essential oils.

To make your own essential oil, crush and pulverize the herb or fruit (skin and all) with a mortar and pestle or in a blender or food processor. Put this in a glass jar and cover (just) with walnut oil (from health food store), or use another cold-pressed oil of your choice and let sit, stirring twice a day, in a warm place for a week. Strain the oil and dilute it with more oil if it is too strong. Squeeze an open vitamin E capsule into the oil to act as a preservative. Add a drop or two to a cleaning formula of your choice.

You can also buy a number of fragrant oils in the grocery store. If you are in need of lemon oil for a furniture polish, try food grade lemon extract. These extracts contain only small quantities of pure oil in a solution of alcohol, but you can use them when you are in a pinch. The alcohol will evaporate. (Make sure you don't buy the imitation.)

Fragrant herbs: There are endless combinations of herbs and flowers to make your cleaning formulas fragrant and fresh smelling. Extracting the perfume to produce an essential oil is discussed above. Other options include making a strong fragrant tea, straining, and using that in a formula, or adding a few drops of a high-quality commercial product such as *Dr. Bronner's Pure Castile Soap* scented with lavender.

Preservatives: There are occasions when you want to preserve an ingredient, such as when you make your own essential oil and you want to save it for a while.

Vitamin E: The cosmetic industry has used vitamin E as a preservative for years. Health food stores carry vitamin E capsules. Simply cut a capsule in half and squeeze contents into the formula.

You can also buy liquid vitamin E with a dropper.

Vitamin C: Buy vitamin C crystals in your health food store and add a little bit to your formula.

Citrus Seed Extract: Citrus seed extract is commonly used in "natural" cosmetics as a preservative. Buy liquid *Paramycocidin* (from NutriCology) and add a drop or two to your mixture.

Saponin: Saponin is a natural ingredient found in some plants. It has intrinsic cleaning and foaming properties. Saponin is present in soapwort's (or Bouncing Bet's) leaves, stems and roots. It is also in yucca root and soapberry root. Before the advent of the petrochemical age, soapwort roots were pulverized, boiled, and used as a clothes bleach.

Solvents: Heavy-duty cleaning sometimes requires the strength of a solvent. Most commercial solvents are highly toxic and considered household hazardous waste. The only natural solvents I know of are *Plant Thinner, Healthy Kleaner,* and *Natural Solvent Spotter.* See Chapter 22 for further information on these products.

Vegetable Glycerin: Often used in cleaning recipes as a stain remover and because it helps oil mix with water.

Vegetable-Oil-Based Liquid Soaps: These are primarily coconut oil based soaps that biodegrade quickly. They are most frequently found in health food stores, usually described as all-purpose or dishwashing soaps.

Vinegar: Cuts grease, dissolves gummy buildup, inhibits mold growth, dissolves mineral accumulation, freshens the air. Recipes in this book that call for vinegar refer to white vinegar.

Animal

Beeswax: A natural, hard wax with a rich, distinctive aroma.

Diatomaceous Earth: Made from skeletons of a prehistoric algae. Used in cleaning because it absorbs oil and water, and has abrasive qualities.

Lanolin: This oil is found in sheep wool. Lanolin is removed at no harm to the sheep. Lanolin is included in cleaners, particularly leather, because of its remarkable penetrating abilities.

Shellac: Lac is a beetle excretion. Mixed with alcohol, lac has become famous as the furniture sealant, shellac.

Sources of Materials

There are primarily three sources for buying the materials you will use for ecological cleaning:

1. The most plentiful sources for natural ingredients are your

health food store and supermarket. Happily, you often have these supplies on hand, in which case you don't even need to buy anything. For example, vinegar and salt. Sometimes the ingredient is even in your refrigerator or garden, such as lemon or rhubarb.

2.   Second, mail order companies are very useful for obtaining some hard-to-find natural ingredients you may need for your recipes. None of these are expensive and I recommend ordering as soon as possible so that they are on hand when you need them. For example, if you want to make a car wax you will need to order a few supplies. To know what ingredients to order check the appropriate chapter's "Clean & Green Commercial Products" listing for a mail-order supplier. You can also ask your local health food store if they can order some of these materials for you. For more details see my suggestions on how to make your shopping list, below.

3.   The third source of supplies are acceptable commercial brands of cleaning products. Health food stores and environmentally oriented mail order suppliers have an increasingly large spectrum of cleaning products to choose from. I highly recommend buying in bulk. Not only do you save time, you also save money and unnecessary packaging. To learn about products you might like to use, see Chapter 22, Commercial Products That Are Safe for Health and Safe for the Earth.

Replacing Commercial Cleaners
    If your cupboards are now bare of synthetic cleaners, pesticides, and paints, you can be happy that you have done a lot for your health and our planet. Helping you replace these items is the purpose of this handbook. I suggest you make a list of indispensable cleaning products. Your list might read like this:
> window cleaner
> dusting aid
> furniture polish
> silver polish
> scouring powder
> dishwashing liquid
> automatic dishwashing detergent
> laundry soap
> fabric softener

    To replace a cleaner, look up the appropriate section in the Table of Contents and find a formula that you like. I suggest making a basic shopping list of ingredients so that when you happen to be at the supermarket or hardware store you can pick up an ingredient even if you don't need it immediately.

CLEAN & GREEN REPLACEMENTS

TYPE	SUPER-MARKET	HEALTH FOOD STORE	HARD-WARE
Window Cleaner	vinegar	Heavenly Horsetail All-Purpose Cleaner	spray bottle
Dusting Aid	vinegar, olive oil	vinegar, olive oil	spray bottle
Furniture Polish	vinegar, food-grade lemon oil	linseed oil, lemon oil	sugar shaker can
Scouring Powder	baking soda		
Silver Polish	baking soda, alum. foil, salt, toothpaste	toothpaste	
Laundry Soap	washing soda	Granny's Power Plus	
Fabric Softener	baking soda		

Accessory Highlights

Cleaning accessories include both familiar implements such as sponges and mops, plus a few special items to make current needs easier, such as spray bottles and vegetable scrub brushes.

Packaging and Containers

Although it is by no means necessary, putting your homemade cleaning products in appropriate containers gives them a nice professional feel and makes cleaning easier. It is also expedient to make up a formula in bulk and save it in a container so that it is ready for the next use.

Commercial Dishwashing Liquid Containers: Before you throw out your dishwashing detergent container, wash the bottle carefully and save it. It is an ideal receptacle for all-purpose liquid soap, especially if you buy it in bulk gallon containers and want something more accessible for dish washing. Don't forget to save the top; the squirt tops save soap.

Glass Jars: Jars, particularly wide-mouth varieties, are great for storing pastes and polishes.

Powder Containers for Abrasive Cleaners: You can buy powder containers (usually used to hold confectioner's sugar) at the hardware store for your homemade abrasive cleaners. I have one that is metal, about 4 inches high, with a handle. The holes in the cans are big enough for the minerals to easily fall through, but if by chance you find one that doesn't work as well, simply enlarge the holes with a nail and hammer. If you want, you can decorate the can. I bought some decals for mine, and it looks pretty anywhere.

Spray Bottles: Window cleaners, furniture polish, and all-purpose cleaners are just a few examples of homemade cleaning agents that work well in a spray bottle. An added bonus of nontoxic cleaning is that there will be no poisons in these bottles, so you can also use them for other projects. Spray bottles are often found in "green thumb" sections of stores.

Aesthetics

A commercial package will have images on it to capture your imagination. A company may use a rose to get you to think the cleaning product will make things smell sweetly, even if it doesn't. We are accustomed to seeing packaging that evokes images, and we can make our own to suit ourselves. I happen to love wild roses, so using a container with rose decals brings pleasant thoughts to me when I clean the bathtub. If you are a cellist, you might like musical notes.

Decals are readily available at specialty stores and are easy to put on spray bottles, glass jars, and powder containers. Better yet, buy nontoxic paint and design your own.

Natural Cleaning Implements

Cellulose Compressed Sponges and French Cotton Sponges: These sponges are flat like a cracker until you get them wet and they pop up into real sponge size. If you are one of the many who dislike using supermarket-brand sponges containing plastic and synthetic resins, you will love these sponges. Not only do they feel better to work with, but they work wonderfully.

Cellulose Sponge Cloth: Made out of pure cellulose, these can be used and reused instead of paper towels. You can even wash them in the washing machine!

Cheesecloth: Furniture polishers, window wipes, and all-purpose cleaning rags are just a few examples of ways to use cheesecloth. It is the cloth of choice for professional polishers. Simply fold it upon itself a number of times.

Natural Bristle Brushes: Although many of the pesticides and herbicides used on our fruits and vegetables are systemic, you can get some spray residue off by washing the vegetables in a safe soap and scrubbing them with a natural bristle vegetable brush. This approach will also help remove the wax (sometimes impregnated with fungicide) that many vegetables are coated with. Natural bristle brushes are also very handy to use when washing dishes. Some brands have replaceable heads, so you don't need to throw out the entire brush when the bristles are beyond salvage.

Natural Sponges: Natural sponges are the skeletons of an aquatic lower invertebrate life form called Porifera. These skeletons are especially effective as sponges because they can both retain and release fluid. Silk sea sponges are one kind of natural sponge that is particularly soft and pliable.

Pure, Natural Latex Work Gloves: These reusable gloves are made from a one hundred percent natural material and therefore they biodegrade completely.

One Hundred Percent Cotton Towels: Instead of using paper towels to wipe things up, buy a dozen or so good, one hundred percent cotton dishtowels. If you have enough cloth towels, you won't mind getting them dirty. Cotton towels are a little like diapers; you can use them over and over again.

Recycled Materials

Old Sponges: Can be washed in vinegar, saved, and used again.

Rags: I have just recently switched from paper towels to rags. When I can, I use a cut-up one hundred percent cotton T-shirt which gets washed and reused. Use something bright and colorful instead of a drab old pajama leg, and you will never switch back to paper towels. Soft rags feel good, wash out so well, and handle so beautifully, you will forever be a convert.

Toothbrushes: Great to clean hard-to-reach areas and toothbrushes don't scratch.

Formulas and Ratios

Calculating the proper amounts of ingredients for your needs may take a little experimentation. For example, in making an olive oil and vinegar polish you can make a "heavy" polish by using as much as three times more olive oil than vinegar. This might be appropriate for very clean, well-dusted and varnished furniture. I prefer a "lighter" polish made by using up to three times more vinegar than olive oil. The vinegar brings up dirt and the olive oil enriches the wood. I make an even more vinegary version for a homemade dusting aid, using ¼ cup vinegar and ½ to 1 teaspoon olive oil. The wood has a wonderful nutty smell for a while, and then the vinegar completely evaporates leaving the wood not only clean but beautiful. The bolder you are in playing with ratios, the more custom-made your cleaning products will be.

Each recipe section of this handbook intentionally has many ingredient alternatives. These formulas are only helpful if they meet your needs, and by having a lot of alternatives, you can adjust a preparation for a specific purpose. For example, you might have a metal pan that needs cleaning but also disinfecting. Borax, from one recipe, is a disinfectant, and lemons, from another, clean tarnish. If you put the two ingredients together, your cleaning problem is solved. Sometimes a cleaning problem is unusually tough, but if you have a choice of ten options, the chances are you will find something that works. If you just can't seem to solve a cleaning problem, see Chapter 17, Stains and Tough Jobs, which has about one hundred stain-removal remedies.

To avoid any mistakes, I suggest testing each new recipe in an inconspicuous place on the item being cleaned. This procedure is particularly important when cleaning stains on rugs, upholstery, and clothing.

Chapter 2

ALL-PURPOSE HOUSEHOLD CLEANERS

One could formulate all-purpose cleaners for days. There are myriad possibilities. As your understanding of nontoxic ingredients grows, you will be able to truly custom make cleaners by mixing to suit your water type, level of wear and tear, and cleaning habits. Refer to Chapter 17, Stains and Tough Jobs, for more ingredient ideas, and to appropriate chapter sections for specific cleaning problems.

☐ **NATURAL INGREDIENT CHOICES**
Supermarket: borax, vinegar, baking soda, washing soda, food-grade essential oils, aromatic herbs, lemon juice
Health Food Store/Mail Order: herbs, vegetable-oil-based liquid soap, essential oils
Garden: aromatic herbs, fragrant flowers
☐ **CLEAN & GREEN COMMERCIAL PRODUCTS**
<u>*Health Food Store/Mail Order*</u>

A.F.M. Enterprises
   Super Clean
   Safety Clean
Auro Organics
   Plant Soap
   Cleansing Emulsion
Biofa
   Household Cleanser
Bon Ami
   Cleaning Powder
   Cleaning Cake
Cal Ben Soap Company
   Pure Soap (bar soap)
   Seafoam Dish Glow

Dr. Bronner's
   Pure Castile Soaps
   Sal Suds
Ecco Bella
   Laundry Booster and
    Whitener
Ecover
   Cream Cleaner
   Dishwashing Liquid
Granny's Old Fashioned
  Products
   Aloe Care

Greenspan
  Friendly Cleaner
Infinity Herbal Products
  Heavenly Horsetail
Jurlique
  Gentle
  Sparkle
Life Tree
  Home-Soap All-Purpose
  Household Cleaner
Livos
  Avi-Soap Concentrate
  Kiros-Alcohol Thinner
  Latis-Soap Concentrate

Naturally Yours
  Spray and Wipe Cleaner
  All Purpose Cleaner
  Degreaser
  Gentle Soap
Pacific Development
Systems
  Handi-Gloves
Simmons Pure Soaps
  Bar soaps
Tropical Soap Company
  Sirena coconut-oil bar
  soaps

Supermarket

Arm & Hammer
  Baking Soda
  Super Washing Soda

Dial Corporation
  20 Mule Team Borax

□ **IMPLEMENTS**: spray bottle, cellulose sponge cloth, natural latex gloves, rags, sponges, silk sea sponges, cotton mops, buckets, and pails

CLEANING TIPS

Spray Bottles: Minerals such as borax, baking soda, and washing soda can clog spray bottles unless they are dissolved completely with very hot water.

Cleaning Fiberglass Warning: In recipes that call for washing soda, substitute baking soda, borax, or nothing. Washing soda can scratch fiberglass. *Ecover's Cream Cleaner* is a safe commercial soap that is suitable for fiberglass.

Hard-Water-Area Cleaning: Hard water has a high mineral content, which leaves a residue. You need more soap for sudsing and more minerals for cleaning. Conversely, soft water has a low mineral count, requiring less soap and minerals. Vinegar and lemon juice are good acids for cleaning mineral buildup. If you live with high-mineral-content water add more vinegar or lemon juice to the recipes. If you really need a solvent, try *Plant Thinner* from Auro Natural Plant Chemistry for the purest citrus solvent.

HEALTH NOTE ON LIQUID SOAPS

Although the chemicals DEA, MEA, and TEA (among others) are suspected of causing the formation of carcinogenic nitrosamines in cosmetics, it is unclear if they do this in soaps. To be on the safe side, I suggest you add a couple of drops of liquid vitamin E (available at your health food store) per ½ gallon to all vegetable-oil-based liquid soaps recommended in this book, to help protect against possible nitrosamine contamination. To make this task easy, just add the vitamin E to the bottle when you open it the first time. You do not have to do this at all if either vitamin E or vitamin C (ascorbic acid) is listed in the ingredients.

Note: All the recipes in this book are formulated using the recommended brands of vegetable-oil-based liquid soaps or detergents mentioned above.

1. FANTASTIC CLEANER
Spray Cleaner:

1 teaspoon borax, ½ teaspoon washing soda, 2 tablespoons vinegar or lemon juice, ¼ to ½ teaspoon vegetable-oil-based liquid soap, 2 cups very hot tap water, spray bottle
• Combine the borax, washing soda, vinegar, and liquid soap in a spray bottle. Add very hot tap water, shaking the bottle gently until the minerals have dissolved. Spray onto the area to be cleaned and wipe off with a sponge, rag or cellulose sponge cloth.

Buckets or Pails:

⅛ cup borax, ⅛ cup washing soda, 1 tablespoon vegetable-oil-based liquid soap, ¼ cup vinegar, 2 gallons hot water
• Place ingredients in a pail. Stir to dissolve. Use with sponges or mops as usual. Because washing soda is slightly caustic, use gloves if the solution will come in contact with your skin. Rinse well.

2. DISINFECTANT, ANTIMILDEW, AND GETS-APPLIANCES-SHINY CLEANER
Spray Cleaner:

1 teaspoon borax, 2 tablespoons vinegar, ¼ teaspoon

vegetable-oil-based liquid soap, 2 cups very hot water, spray bottle
- Follow directions for Fantastic Cleaner.

Buckets or Pails:
¼ cup borax, ¼ cup vinegar, 2 gallons hot water
- Follow directions for Fantastic Cleaner.

3. GREASE CUTTER
Spray Cleaner:
½ teaspoon washing soda, 2 tablespoons vinegar, ¼ teaspoon vegetable-oil-based liquid soap, 2 cups very hot tap water, spray bottle
- Follow directions for Fantastic Cleaner.

Buckets or Pails:
¼ cup washing soda, ¼ cup vinegar, 1 tablespoon vegetable-oil-based liquid soap, 2 gallons hot water
- Follow directions for Fantastic Cleaner.

4. THIS IS THE BEEF
Spray Cleaner:
½ teaspoon baking soda, ½ teaspoon borax, ½ teaspoon washing soda, 2 tablespoons vinegar, ½ teaspoon vegetable-oil-based liquid soap, 2 cups very hot water, spray bottle
- Follow directions for Fantastic Cleaner.

Buckets or Pails:
⅛ cup baking soda, borax and washing soda, 1 tablespoon vegetable-oil-based liquid soap, 2 gallons hot water
- Follow directions for Fantastic Cleaner.

5. PLAIN AND SIMPLE
Spray Cleaner:
1 teaspoon borax, 2 tablespoons vinegar, 2 cups hot water, spray bottle
- Follow directions for Fantastic Cleaner.

Buckets or Pails:
¼ cup borax, ¼ cup vinegar, 2 gallons hot water
- Follow directions for Fantastic Cleaner.

6. VEGETABLE-OIL-BASED SOAP
⅛ cup vegetable-oil-based soap, ½ cup cooled fragrant herb tea, 2 gallons hot water

• Follow directions for Buckets or Pails under Fantastic Cleaner.

7. BASEBOARDS AND OTHER THINGS
Spray Cleaner:
   1 tablespoon cornstarch, ¼ cup vinegar, 2 cups hot water
   • Dissolve cornstarch in spray bottle with hot water. Add vinegar. Spray and wipe.
Buckets or Pails:
   ¼ cup cornstarch, ½ cup vinegar, 1 gallon water
   • Follow directions for Fantastic Cleaner.

8. FRESH AND FRAGRANT ALL PURPOSE CLEANER
   Add a few drops of an essential oil to the recipe of your choice. An alternative to essential oils is to add a fragrant herb or flower tea (½ cup or so), or a few drops of one of Dr. Bronner's scented pure castile soaps. Adding a scent to a cleaning formula can help make a room feel light and airy. To find out more about essential oils see Chapter 1, Ecological Cleaning.

Chapter 3

SCOURING POWDERS AND SOFT SCRUBBERS

Many of us remember growing up with scouring powder and a sponge under both the kitchen and bathroom sinks. The scouring powder came in a convenient shaker can, ready to use at any time. There is no reason that you can't have exactly the same convenience, nontoxically. Simply buy a couple of confectioner's sugar shaker cans, fill with your *Clean & Green* formula, and place along with a sponge under the sinks. You can do the same with your own *Clean & Green* soft scrubbers stored in recycled glass jars.

□ **NATURAL INGREDIENT CHOICES**
Supermarket: baking soda, borax, washing soda, food-grade essential oils, aromatic herbs
Health Food Store/Mail Order: vegetable-oil-based liquid soap, pumice stone powder, sodium perborate, chalk
Garden: aromatic herbs, fragrant flowers
Hardware store: soft blackboard chalk
□ **CLEAN & GREEN COMMERCIAL PRODUCTS**
Health Food Store/Mail Order

Bon Ami
   Cleaning Powder
Ecco Bella
   Laundry Booster and
   Whitener
Ecover
   Cream Cleaner

Karen's Nontoxic Products
   Pumice stone stick
Wood Finishing Supply
Company
   Pumice Powder

Supermarket

Arm & Hammer
  Baking Soda
  Super Washing Soda
Bon Ami
  Cleaning Powder

Dial Corporation
  20 Mule Team Borax

See Chapter 2, All Purpose Cleaners, for brand names of vegetable-oil-based liquid soaps.

□ **ECO-BUY:** Natural minerals (below) make economical scouring powders.

□ **IMPLEMENTS:** Confectioners sugar shakers are very handy, with holes usually big enough for minerals. Natural silk sea sponges are nice for applying scouring powders, as are cellulose sponges, cellulose sponge cloths, and French cotton sponges. Avoid abrasive pads that can scratch.

INGREDIENT HIGHLIGHTS

Baking Soda: The virtues of baking soda cannot be emphasized enough. It's amazing in its powers to eat up chemicals and odors. I use it instead of any other commercial scouring powder. I wash everything with it. If you have used it as a cleaner and weren't satisfied, it is probably because you weren't using enough.

Mineral Magic: Custom-make your own cleaners by mixing different combinations of the following minerals. The bigger the job, the more mineral you need, and the more you need to rinse.

Baking soda: absorbs odors, deodorizes, mild abrasive
Borax: disinfects, deodorizes, inhibits mold
Chalk: mild non-abrasive cleaner
Pumice: removes stains, polishes
Salt: mild abrasive
Sodium perborate: bleaching agent
Washing soda: cuts grease, best for tough jobs
Zeolite: ion exchanger, absorbs pollutants

CLEANING TIPS

Rinse Well. Minerals can leave a residue, particularly in a high concentration.

Cleaning Fiberglass: In recipes that call for washing soda, substitute baking soda or borax. Washing soda scratches fiberglass. A commercial product that can be recommended for cleaning fiberglass is *Ecover's Cream Cleaner*.

Hard-Water-Area Cleaning Tip: Vinegar and lemon juice help clean mineral buildup.

Mold Protection for Tiling: If you have mold in the grout, add a bit of borax to the baking soda. The mold will be killed and the bathroom disinfected.

Chalk: When a recipe calls for chalk, I mean soft, white blackboard chalk. The old fashioned term for chalk is whiting and you may be able to find this powder in specialty stores. If not, simply place chalk sticks into a paper bag, roll the bag up, and with a hammer pulverize the chalk to a powder. Pour the powder into a bowl and grind any remaining lumps with a pestle or rock.

Storage: Soft scrubbers can be made in bulk and stored in glass jars. Stir before reusing. If they dry out, add a little more soap.

Scouring Powders

9. BAKING SODA SCOURING POWDER
½ cup baking soda

• Place the baking soda in a bowl. Dampen a sponge, scoop the baking soda up with it, and rub onto the surface to be cleaned. Let the baking soda "rest" on the surface for a while to absorb odors, then rinse thoroughly. (If you want to use a shaker, place the baking soda in the can and shake onto the surface. Rub in with a damp sponge.)

10. BAKING SODA AND ROSE PETALS
1 cup baking soda, fragrant herb or flower petals

• You can personalize your scouring powder by adding an aromatic herb or flower. Put the ingredients into a blender and run the machine until the fragrance has infused the powder. Add a drop or two of water if you need a more malleable consistency. Pour the scented baking soda onto the surface to be cleaned and scrub with a sponge. Rinse well. If you have any left over, leave it in an open dish as an unusual version of potpourri. People with pollen allergies should choose their herb carefully.

11. DISINFECTANT SCOURING POWDER
1 cup baking soda, ¼ cup borax

• Follow directions for Baking Soda Scouring Powder. Rinse well.

12. TO CUT GREASE

1 cup baking soda, ¼ cup washing soda

• Follow directions for Baking Soda Scouring Powder. Rinse well.

13. SCOURING POWDER THAT WHITENS TOO

½ cup baking soda, 3 tablespoons sodium perborate

• Follow directions for Baking Soda Scouring Powder. Rinse well.

14. ANTIMILDEW SCOURING POWDER

⅔ cup baking soda, ⅓ cup borax

• Follow directions for Baking Soda Scouring Powder. If the spot being cleaned is really moldy, you might try scouring the area with straight borax, but before you rinse, let the mineral "rest" on the area for a couple of hours. Rinse well.

15. PUMICE STICK FOR PORCELAIN STAINS

pumice stick, water

• Place the pumice stick under running water until it is saturated. Holding one end of the stick, rub difficult-to-remove stains until they are gone. Rinse.

16. ALUM AND CREAM OF TARTAR FOR PORCELAIN

2 or 3 tablespoons cream of tartar, 2 or 3 tablespoons alum, water

• Lightly dampen porcelain. Sprinkle minerals onto stain. Let rest for a few hours before rubbing with a sponge and rinsing.

17. SALT SCRUBBER

iodized or uniodized salt

• Pour salt directly onto the area to be cleaned and rub with a sponge. Rinse well.

Soft Scrubbers

18. CREAMY-LIKE-FROSTING SOFT SCRUBBER

¼ cup baking soda, enough vegetable-oil-based liquid soap to make a paste

• Place the baking soda in a bowl and stir in liquid soap, stirring as you add, until you have a rich, creamy, texture. Scoop the mixture onto a sponge, wash the surface, and rinse thoroughly.

This recipe makes a soft scrubber that is so luxuriantly creamy, I recommend not putting it any place someone could mistake it for frosting.

19. NON-ABRASIVE SOFT SCRUBBER

¼ cup powdered chalk, enough vegetable-oil-based liquid soap to make a paste

• Follow directions for Creamy-Like-Frosting Soft Scrubber. See Cleaning Tips, above, on making powdered chalk.

20. DISINFECTANT/MOLD ELIMINATING SOFT SCRUBBER

¼ borax, enough vegetable-oil-based liquid soap to make a paste

• Follow directions for Creamy-Like-Frosting Soft Scrubber. The texture for this recipe is slightly grainy.

21. HEAVY DUTY CLEANING SOFT SCRUBBER

¼ cup washing soda, enough vegetable-oil-based liquid soap to make a paste

• Follow directions for Creamy-Like-Frosting Soft Scrubber. Wear gloves when applying.

22. PUMICE FOR PORCELAIN SOFT SCRUBBER

¼ cup pumice, enough vegetable-oil-based liquid soap to make a paste

• Follow directions for Creamy-Like-Frosting Soft Scrubber. Remember, pumice powder should not be inhaled.

Chapter 4

DISINFECTANTS AND MOLD CLEANERS

The word *disinfect* means to cleanse or rid of pathogenic microorganisms. For a substance to be classified as a disinfectant with the federal government, it needs to go through expensive and elaborate tests. The ingredients I recommend as disinfectants have not necessarily been through such stringent testing. Borax is a good example of a remarkable material that is considered so effective at killing microorganisms and fungi that even hospitals use it. However, it is not technically classified as a disinfectant. Legend and folklore have their place, also, and I think that herbs that for millennia have been used for cleansing wounds and antisepsis are worth knowing about and experimenting with.

Mold can be removed or prevented by:
1. Exposure to sunlight.
2. Eliminating moisture through fresh dry air, heat, a dehumidifier, and removing the source of the problem.
3. The application of antifungal agents, as listed below.

INGREDIENT HIGHLIGHTS
Australian Tea Tree Oil: This essential oil is an excellent broad spectrum fungicide and bactericide.
Borax: Borax has been proven to be very effective at killing microorganisms and also has natural antifungal properties.
Citrus Seed Extract: Citrus fruit has stunning cleaning properties. Grapefruit seed extract is considered antifungal, antimicrobial and antibacterial. It is used as a preservative in cosmetics. The only pure form of citrus seed extract I know of is paramycocidin, sold as a dietary supplement for killing internal parasites.
Pine Oil: The reason pine oil is in so many commercial cleaners is that it has disinfectant properties. Karen's Nontoxic Products (see Mail Order Suppliers) carries a pure pine oil. Pine oil is a very allergenic substance, so test carefully before using.
Zeolite: Zeolite absorbs bacteria, mold, and other pollutants from

the air. Buy "breather bags" of *Odor-Fresh Zeolite* from G&W supply and place in the refrigerator or other problematic places. *Odor-Fresh Zeolite* is also available in powder form to add to cleaners. Zeolite is a one hundred percent nontoxic, naturally occurring mineral found near volcanic activity. What makes zeolite unusual is that it is the only mineral that is an ion exchanger in its natural state, which means it naturally absorbs pollutants from the air. It absorbs ammonia from water and can clear an entire house of severe smoke damage in three days. To do this without the use of toxic chemicals or ozone generators is almost unheard of. It is claimed that zeolite will also extend the shelf life of fruit and vegetables for as long as three weeks. Zeolite itself is available in four dimensions from as fine as talc (which is dusty) to as large as pellets of ⅜ inch in diameter. Zeolite also "de-sorbs" - just put it in the sun and the fumes will disappear and you can use it again and again.

Disinfectants

□ **NATURAL INGREDIENT CHOICES**
Supermarket: borax, vinegar, herbs
Health Food Store/Mail Order: herbs, pine oil
Garden: aromatic herbs, fragrant flowers
□ **CLEAN & GREEN COMMERCIAL PRODUCTS**
Health Food Store/Mail Order

A.F.M. Enterprises	Karen's Nontoxic Products
Safety Clean	Pine Oil

Supermarket

Dial Corporation
20 Mule Team Borax
□ **ECO-BUY:** Ounce for ounce, borax is an extremely cheap disinfectant.

---

23. BORAX CLEANER
½ cup borax, ¼ cup vinegar, 2 gallons hot water
• Dissolve borax, along with vinegar, in very hot water and wash the surface using a mop or sponge, as appropriate. To make a stronger disinfectant solution, add more borax. Rinse thoroughly.

24. BORAX AND LAVENDER
• Follow directions for Borax Cleaner but add a few drops of the essential oil of the antiseptic fragrant herb lavender.

25. BORAX SPRAY

1 teaspoon borax, 3 tablespoons vinegar, 2 cups very hot tap water, spray bottle

• Combine the ingredients in the spray bottle and dissolve completely with hot water. Spray the area, let it "rest" for a while, and then rinse if the borax has left a residue.

26. AUSTRALIA'S FABULOUS FUNGICIDE

2 teaspoons Australian tea tree oil, 2 cups water, spray bottle

• I used this formula on a bureau that always smelled musty. I sprayed every part of the bureau. I didn't rinse. The tea tree oil eliminated the musty smell and the strong odor of this essential oil dissipated after a few days.

27. PINE OIL

a few drops pine oil, cleaning formula of your choice

• Pine oil is commonly used in household cleaners because of its disinfectant properties. Combine the oil with the cleaning formula of your choice or water, and use on the surface being disinfected. Rinse.

Note: Pine oil is a highly allergenic substance, so use the smallest amount possible, keep out of reach of children, and test carefully.

28. ANTISEPTIC HERBS

3 or 4 tablespoons dried, pulverized heather, myrrh, sage, peppermint, rosemary, eucalyptus, wormwood, rue, lavender, thyme, mint, or 1 teaspoon ground clove, tea pot, 2 cups boiling water, spray bottle

• Make a strong tea by placing one of the herbs listed in the tea pot, filling the pot with boiling water, and steeping for ten minutes or so. Cool and strain into the spray bottle. Spray where you need antisepsis (pet odors, bathrooms, etc.). Rinse well.

Mold Cleaners

□ **NATURAL INGREDIENT CHOICES**
Supermarket: borax, vinegar, deodorant-free kitty litter, cornstarch
Health Food Store/Mail Order: zeolite, Australian tea tree oil
Hardware Store: heat lamp, light bulb, lime

☐ CLEAN & GREEN COMMERCIAL PRODUCTS

Health Food Store/Mail Order

A.F.M. Enterprises
  X158 Mildew Control
Desert Essence
  Australian Tea Tree Oil
G&W Supply
  Odor-Fresh Zeolite

Naturally Yours
  Mold and Mildew
  Remover
NutriCology
  Liquid Paramycocidin

Supermarket

Dial Corporation
  20 Mule Team Borax

---

29. BORAX DISINFECTANT AND MOLD KILLER

1 teaspoon to ¼ cup borax, up to 2 cups hot tap water
• Place the borax in a container and dissolve completely in hot tap water. Saturate a sponge with the mixture and wash the moldy area. If really moldy, use an even higher concentration of borax and/or leave the solution on for a few hours or overnight, then rinse well. The more borax, the more residue to rinse off, but borax really works. This can even be used to clean plaster walls that have been penetrated by mold by using an almost straight borax paste. Leave the borax on the walls for a number of days and when it is completely dry, vacuum up the powder.

30. VINEGAR MOLD KILLER

full-strength vinegar
• Saturate a sponge with vinegar and scrub the moldy area. Rinse well.

31. NO MOISTURE - NO MOLD

deodorant-free kitty litter
• Kitty litter is used here for its moisture absorbent properties. Place bowls of the material in damp areas. Replace with fresh kitty litter every week or so.

32. NO MOISTURE - NO MOLD II

pure lime
• Put buckets of lime in the basement to absorb moisture. Don't use chlorinated lime.

33. ELIMINATE MILDEW FROM BOOKS

cornstarch

• Place cornstarch in a shaker and sprinkle evenly onto pages. Let sit for a few hours to absorb any moisture. Wipe clean.

34. ZEOLITE FOR MOLD AND BACTERIA

2 or 3 "breather bags" of zeolite

• Place zeolite bags in moldy, damp rooms. Zeolite will absorb the offending odors, especially if you use enough "breather bags." On warm sunny days put the "breather bags" in the sun to cleanse the zeolite, then put them back in the damp room.

35. GRAPEFRUIT SEED EXTRACT

10+ drops of liquid Paramycocidin, 2 cups water, spray bottle

• You may need more or less Paramycocidin, depending on the degree of mold infestation. Don't rinse; let it stay in place and continue to do its work. For information on how to obtain Paramycocidin, see NutriCology in Chapter 23.

Chapter 5

AIR FRESHENERS AND ODOR REMOVAL

Natural air fresheners can make a room feel vibrant and fresh; chemical air fresheners interfere with your smelling ability and in the end just introduce a load of synthetics into your home and body. Adding some lemon juice to a cleaning formula or placing a vase of fresh flowers in a stuffy location are simple, pleasant alternatives to store bought air fresheners. Here are my favorite ways of clearing the air.

☐ NATURAL INGREDIENT CHOICES
Supermarket: citrus fruit, baking soda, natural kitty litter, dry mustard, white bread, borax, washing soda, vanilla and other flavoring extracts
Health Food Store/Mail Order: washing soda, zeolite
Plant Store: English ivy, aloe vera, fig tree, spider plant, potted chrysanthemum
Garden: fragrant herbs

☐ CLEAN & GREEN COMMERCIAL PRODUCTS

Health Food Store/Mail Order

A.F.M. Enterprises
   Safety Clean
   Super Clean
Ecco Bella
   Laundry Booster and
   Whitener (calcium
   carbonate)

Arm & Hammer
   Baking Soda
   Super Washing Soda

G&W Supply
   Odor-Fresh Zeolite
Mia Rose
   Air Therapy
Wysong
   Citressence

Supermarket

Dial Corporation
   20 Mule Team Borax

Note: For cleaning the air of chemical odors, see Chapter 19, Cleaning Up Chemicals.

36. ZEOLITE
3 or 4 breather bags, per room, of Odor-Fresh Zeolite
* Place breather bags in areas where you want odor removed.

37. FRAGRANT HERB AIR FRESHENER
fragrant herb or flower of your choice, water
* Make your own fragrant herb infusion by pulverizing the herb or leaves of your choice in a blender or with a mortar and pestle. Put the pulp in the bottom of a tea pot and cover with a cup or two of boiling water. Once cooled, put the pulp and water into a glass jar and let steep for four or five days. Strain. Pour liquid into a small bowl or cup and place in different parts of the house. For an air freshener spray, put some of the liquid into a spray bottle and spray around locations you want smelling fresh.

Alternative 1: Put cloves, cinnamon sticks, or other item of choice in a pan with enough water to simmer for an hour or two. Let the mixture gently bubble away.

Alternative 2: Place potpourri or flowers in appropriate places.

Alternative 3: Put a teaspoon of vanilla or other flavoring extract in an uncovered container and place in appropriate places.

38. CITRUS AIR FRESHENER
a few slices of lemon, grapefruit or orange, pot, water
* Place the citrus in a pot with enough water to cover. Simmer gently in the open pot for an hour or so. If you use aluminum pots, this will clean them, too.

39. IT ALWAYS WORKS: BAKING SODA
baking soda
* Place the baking soda in any open container of your choice and set in small enclosed areas like closets.

40. PLANTS CAN CLEAN THE AIR
* NASA has discovered that plants can absorb and neutralize air pollution. This works best in an enclosed space. The more plants the better. The most effective plants include: aloe vera (many toxins), English ivy (benzene), fig tree (formaldehyde), potted chrysanthemum (general, many toxins), spider plants (formaldehyde).

41. FIREPLACE CLEANER
¼ cup or less washing soda, 2 gallons hot tap water
• Dissolve washing soda in hot water and scrub on stonework with a brush or mild abrasive pad such as a supermarket green pad. The more washing soda, the more rinsing needed.

42. WHITE BREAD FOR REFRIGERATOR ODORS
2 or 3 slices white bread
• Put a few pieces of bread in the back of the refrigerator to eat up smells.

43. ZEOLITE FOR FRIDGE ODORS
a small "breather bag" of zeolite
• Placed in refrigerator, zeolite will eat up odors and is said to also extend the shelf life of vegetables.

44. ZESTY LEMON FOR FOOD SMELLS
1 lemon, sliced
• Place sliced lemon in a dish near the food preparation area.

45. CUTTING BOARD ODOR-BE-GONE
sliced lemon
• Rub the cut lemon onto the washed cutting board to eliminate lingering smells.

46. BUTCHERBLOCK CLEANER
½ cup vinegar, ½ cup water
• Wash butcherblock with mixture. Rinse.

47. CUTTING BOARD ODOR ELIMINATOR
baking soda
• Sprinkle baking soda onto the cutting board and scrub in with a sponge. Rinse well.

48. FOOD SMELL NEUTRALIZER
celery
• Cut a piece of celery and rub the cut part over places that have had contact with odiferous food. Rinse.

49. PRETTY FISHY
2 tablespoons dry mustard, ½ cup water

• Wash the areas that smell like fish with this mixture, or add 2 tablespoons of dry mustard to the water in which you are washing the fish dishes.

50. BAKING SODA AND LEMON SPRAY
1 teaspoon baking soda, 1 teaspoon lemon juice, 2 cups hot tap water, spray bottle
• Dissolve the ingredients completely in hot water. Place in a spray bottle. Spray as you would an air freshener.

51. BORAX DEODORIZER
¼ to ½ cup borax, 2 gallons hot tap water
• Place the borax in a bucket and dissolve in hot water. Wash areas that need deodorizing.

52. KITTY LITTER FOR SMELLS
1 bag undeodorized kitty litter
• Put the kitty litter in bowls and place in areas where you want to be rid of smells.

53. VINEGAR STRAIGHT UP
¼ cup vinegar
• Place the vinegar in a cup and set in areas of the home where you need air freshening. I have heard that this even works to eliminate cigarette smoke.

Chapter 6

THE KITCHEN

Dishes and Dishwashers

□ **NATURAL INGREDIENT CHOICES**
Supermarket: vinegar, washing soda, borax, baking soda
Health Food Store: vegetable-oil-based liquid soap, bar soap
□ **CLEAN & GREEN COMMERCIAL PRODUCTS**

Health Food Store/Mail Order

Biofa
   Dishwashing Liquid
Ecover
   Dishwashing Liquid
Cal Ben Soap Company
   Pure Soap (bar soap)
   Seafoam Dish Glow
Chef's Soap
   Chef's Soap (bar soap)
Granny's Old Fashioned
 Products
   Aloe Care
Infinity Herbal Products
   Heavenly Horsetail All
   Purpose Cleaner

Jurlique
   Sparkle
Life Tree
   Automatic Dishwashing
   Detergent
   Premium Dishwashing
   Liquid
Naturally Yours
   Degreaser
   Dishwashing Detergent
Simmons Pure Soaps
   Castile and vegetarian
   bar soaps
Tropical Soap Company
   Sirena coconut-oil bar
   soaps

Supermarket

Arm & Hammer
   Baking Soda
   Super Washing Soda

Dial Corporation
   20 Mule Team Borax

□ **IMPLEMENTS**
Cotton Dishtowels: Instead of using paper towels to wipe things up, buy a dozen or so good quality one hundred percent cotton dishtowels.

Hand Dishwashing Equipment: Besides an assortment of pure cellulose sponges, cellulose sponge cloths, cotton compressed sponges, and even silk sea sponges, I highly recommend a "Lola" to scrub off food residue. Lolas are wood-handled dishwashing brushes with replaceable heads.

Warning: Washing soda should not be used on aluminum.

HEALTH NOTE ON LIQUID SOAPS
See Health Note on Liquid Soaps in Chapter 2, All-Purpose Household Cleaners.

DISHWASHERS
The only natural, effective, and nontoxic material I know of that works well in a dishwasher is sodium hexametaphosphate. While a superior choice over supermarket dishwashing machine detergents because it is nontoxic, its phosphate status eliminates it as an acceptable choice for the environment. One very good, pure, commercial product is made by Life Tree.

Dishwashing By Hand

54. LIQUID DISH WASHING SOAP
vegetable-oil-based liquid soap
• There are a number of acceptable vegetable-oil-based liquid soaps (listed above) that are very good for washing dishes. These biodegrade quickly. Use just like you would a supermarket brand. I recommend putting a couple of drops of vitamin E into the bottle when you first open it as described in Chapter 2.

55. HOMEMADE LIQUID DISH SOAP
unperfumed bar of soap, water
• Make your own soap flakes by grating a bar of pure soap into a sauce pan. Cover the gratings with water and simmer over a low heat until they melt. This takes a while; for a full bar of soap allow two hours. Pour into a suitable container and use as you would any liquid dishwashing soap. For this recipe it is particularly important to obtain chemically clean soap so that impurities don't volatilize into unhealthy gasses. Fortunately, many companies list ingredients.

56. FRENCH TRADITIONAL
unscented plain soap
• Some kitchen stores carry a French soap that comes on a wooden rack for the express purpose of washing dishes. It is used by rubbing a damp sponge or pad over the bar. It is sometimes called Chef's Soap. If you can't find it, check during the holiday season.

57. SPOT AWAY
1 or 2 tablespoons vinegar
• Add vinegar to the dish water.

58. SPOT AWAY II
1 or 2 teaspoons borax
• Add borax to the rinse water.

59. POTS AND PANS
1 tablespoon washing soda, 1 tablespoon vegetable-oil-based liquid soap, water
• Put the washing soda in the dishwashing basin with the vegetable-oil-based soap. Add hot water to dissolve the mineral, and wash your pots.

60. BURNED-ON-FOOD SOAK
3 tablespoons washing soda, dishpan of hot water
• Dissolve the washing soda in the hot water and immerse the pots and pans. Soak until the grease, etc., lifts off. Wash with soap and water. Rinse well.

61. VINEGAR CUTS GREASE
3 tablespoons vinegar
• Add vinegar along with your liquid dish soap for each batch of dishes.

62. TEFLON CLEANER
3 tablespoons baking soda, 2 or 3 lemon slices, water
• Put ingredients in the dirty teflon pot or put the teflon item to be cleaned along with the ingredients in a big pot. Add water to cover the stains. Simmer on the stove until clean.

Garbage Disposal

63. CITRUS FRESHENER FOR THE GARBAGE DISPOSAL

citrus (lemons, grapefruit, or oranges), cut in slices
• Grind the cut citrus, rinds and all, in the disposal. Rinse with hot water.

64. GARBAGE DISPOSAL CLEANER

¼ cup borax
• Every two weeks or so pour some borax into the garbage disposal as a disinfectant.

65. GARBAGE DEODORIZATION

• Each time you add garbage to your garbage pail, sprinkle a handful of baking soda on top to help control malodor.

Drain And Septic Cleaners

☐ **NATURAL INGREDIENT CHOICES**
Supermarket: baking soda, cream of tartar, salt, washing soda, lemon juice
☐ **CLEAN & GREEN COMMERCIAL PRODUCTS**
Supermarket
Arm & Hammer
  Baking Soda
  Super Washing Soda

---

66. DRAIN MAINTENANCE

½ cup baking soda, 3 cups boiling water
• Pour the baking soda down the drain and follow with the boiling water. Let the baking soda and boiling water gurgle and bubble for a while before rinsing with hot tap water.

67. BAKING SODA AND VINEGAR

½ cup baking soda, ½ cup vinegar or lemon juice
• The chemical interaction between baking soda and acid (vinegar or lemon juice) will make a lot of noise - just so you know! Pour the baking soda into the drain first and follow with the acid ingredient. Let the mixture "rest" (it will sound like Mt. Vesuvius) for 15 minutes before rinsing with hot tap water.

68. WASHING SODA EACH WEEK KEEPS THE PLUMBER ASLEEP

¼ cup washing soda

• Pour some washing soda down the drain each week, then rinse with hot tap water.

69. JUST VINEGAR

¼ cup vinegar

• Pour vinegar down the drain.

Chapter 7

GLASS CLEANERS

□ **NATURAL INGREDIENT CHOICES**
Supermarket: vinegar, lemon, salt, cornstarch, tea, mustard, cola, washing soda
Health Food Store/Mail Order: vegetable glycerin, vegetable-oil-based liquid soap
□ **ECO-BUY:** For zero wiper cost use old newspapers. (Petroleum-sensitive people should use with caution.)

Note: These recipes were formulated with the vegetable-oil-based liquid soaps recommended in Chapter 2, All-Purpose Cleaners. Less pure soaps may not work as well.

Windows

70. WINDOW WIPES
  • Make your own "window wipes" by pouring the ingredients of any of the following formulas onto a clean rag or natural cellulose sponge cloth. Store in a covered glass jar for reuse.

71. THE BEST WINDOW WASH
  ¼ to ½ teaspoon vegetable-oil-based liquid soap, 3 tablespoons vinegar, 2 cups water, spray bottle
  • This formula is perfect. Just put all the ingredients in the spray bottle, shake it up a bit, and use just as you would a commercial brand. I particularly recommend *Heavenly Horsetail All-Purpose Cleaner* for the soap in this recipe.

72. FOR PARTICULARLY GREASY WINDOWS
  ¼ teaspoon vegetable-oil-based liquid soap, 3 tablespoons vinegar, ¼ teaspoon washing soda, 2 cups water
  • Follow directions for The Best Window Wash.

73. THIS REALLY WORKS, FOLKS: CORNSTARCH
3 tablespoons cornstarch, ½ cup water
• Put the ingredients in a bowl and mix well. Dab a soft, 100% cotton rag into the cornstarch mixture and rub on the window. At first there will be a film, but as you keep rubbing, the window will become so clean you feel you could walk through it.

74. VINEGAR
⅛ cup vinegar, 1 cup water, spray bottle
• Use as you would any window-cleaning spray.

75. LEMONS
juice from 2 lemons
• Saturate a rag with the lemon juice and rub on the windows.

76. CLEANING WITH THE SUNDAY PAPER
newspapers
• Rub the windows clean with newspaper.
Note: The chemically sensitive should use this approach with caution.

Miscellaneous Glass Cleaners

77. GLASS CLEANER
3 tablespoons salt, ½ cup cold water
• This recipe is useful for jobs where you require an abrasive. Put ingredients in a bowl. Saturate a sponge with the mixture, making sure it is coated with salt, and wipe the surface clean. Rinse thoroughly.

78. TEA AND VINEGAR
1 cup strong black tea (cooled), 3 tablespoons vinegar
• Put tea and vinegar in a spray bottle and use as you would any cleaner. I especially like this on mirrors.

79. TEA LEAVES AND LEMON JUICE FOR GLASS AND CERAMIC VASES
1 tablespoon leaf tea or two tea bags, boiling water, ¼ to ½ cup lemon juice
• Make two cups of strong tea. Let cool. Pour the cooled tea and lemon juice into the vase to be cleaned. Let sit for a couple of hours, then scrub with a dishwashing brush, rinse, and dry.

80. COLA GREASE CUTTER (REGULAR OR DIET)
½ cup cola
• Use cola for dissolving greasy finger marks on the windows. Saturate a sponge or rag with the soda and wipe windows clean.

81. MIRROR FOG PREVENTION
½ teaspoon vegetable glycerin, water
• Saturate a sponge or rag with water, then dab on the glycerin. Rub the solution onto the mirror before taking a shower.

82. MINERAL BUILD-UP ON GLASS
1 teaspoon alum, ¼ cup acidic liquid like vinegar or lemon juice
• Mix the ingredients in a bowl. Saturate a "window wipe" with the mixture and rub on the location needing cleaning.

83. SALT FOR VASES
2 or 3 tablespoons salt
• Using a damp natural bristle brush, rub the salt onto vase stains and scrub until clean. Rinse well.

Chapter 8

APPLIANCES AND OVENS

Ovens

Although cleaning ovens isn't fun under the best of circumstances, it can be made easier. You most certainly want to avoid commercial oven cleaners. Not only are they toxic to inhale, if you don't clean up all the residue, when you heat the oven the gaseous fumes will permeate the air and possibly the food. Note, too, that oven cleaners are considered one of the worst household environmental pollutants.

☐ **NATURAL INGREDIENT CHOICES**
Supermarket: baking soda, salt, washing soda, borax
Health Food Store: vegetable-oil-based liquid soap
☐ **CLEAN & GREEN COMMERCIAL PRODUCTS**
<u>Health Food Store/Mail Order</u>
Ecover
    Cream Cleaner
<u>Supermarket</u>
Bon Ami
    Cleaning Powder

INGREDIENT HIGHLIGHTS
Mineral Magic: Custom-make your own cleaners by mixing any of the following minerals. You can use them by themselves or add them to a recipe of your choice. High concentrations of dissolved minerals can leave a residue, so rinse well. The bigger the job, the more mineral you need, and the more you need to rinse.
    Baking soda: absorbs odors, deodorizes, mild abrasive
    Borax: disinfects and deodorizes
    Salt: mild abrasive
    Washing soda: cuts grease
    Zeolite: ion exchanger, absorbs pollutants

CLEANING TIPS

Homemade "Easy Wipes": Buy a few natural cellulose sponge cloths and saturate them with the cleaning formula of your choice. Save them in a closed glass jar until ready to use again.

84. "BELIEVE IT OR NOT" - THE BEST AND EFFORTLESS OVEN CLEANER

baking soda, water, a squirt or two of liquid soap

• Sprinkle water generously over the bottom of the oven, then cover the grime with baking soda. Sprinkle some more water on top of the baking soda. If you let it sit overnight you can effortlessly wipe up the grease the next morning. I suggest using a mild abrasive pad such as a supermarket green pad to help loosen stubborn spills. When you have cleaned up all the mess, dab a little bit of recommended vegetable-oil-based soap on a sponge and wash the sides, top, and inside of the door, as well as any grease or baking soda residue on the bottom. Rinse thoroughly to remove all baking soda.

85. TOUGH-JOB OVEN CLEANER

1 small box baking soda, ¼ cup washing soda

• Follow the directions for "Believe It or Not," but add washing soda, particularly to burnt-on areas. Washing soda will help cut the grease, but it requires a lot of rinsing.

86. SALT

salt, hot water

• Pour salt and hot water over grease and grime. Let sit for a couple of hours or overnight before scrubbing with a mild abrasive pad like a supermarket green pad. Pour salt directly onto the grease when freshly spilled and come back to it later.

87. MICROWAVE CLEANER AND DEODORIZER

3 or 4 tablespoons baking soda, water

• Make a paste with baking soda and water. Use on a sponge to wash the interior and around the doors. Rinse thoroughly.

88. SOAP-AND-WATER MICROWAVE CLEANER

vegetable-oil-based liquid soap, water

• Dilute soap and use a sponge to wash oven interior and around the door. Rinse thoroughly.

Appliances

89. BORAX-AND-VINEGAR SPRAY
1 teaspoon borax, 3 tablespoons vinegar, 2 cups hot tap water, spray bottle
- Place the ingredients in a spray bottle, shake to mix and dissolve the borax, and spray on appliances. Wipe off with a soft cloth or sponge.

90. SHINY APPLIANCES
1 teaspoon borax, ¼ to ½ teaspoon vegetable-oil-based liquid soap, 3 tablespoons vinegar, 2 cups hot water, spray bottle
- Follow directions for Borax-and-Vinegar Spray.

91. GREASE CUTTER
½ teaspoon washing soda, 1 teaspoon borax, ¼ to ½ teaspoon vegetable-oil-based liquid soap, 3 tablespoons vinegar, 2 cups hot water, spray bottle
- Follow directions for Borax-and-Vinegar Spray but let "rest" for a while before rinsing to loosen the greasy spots.

92. HARD-WATER SPOTS
1 tablespoon alum, 3 tablespoons lemon juice, 2 cups hot water
- Follow directions for Borax-and-Vinegar Spray but leave the spray on a while to loosen scale and mineral deposits. Rinse thoroughly.

93. JUST PLAIN SOAP AND WATER
¼ to ½ teaspoon vegetable-oil-based liquid soap, 2 cups water
- Mix the ingredients in a spray bottle or bowl. Either saturate a sponge with the soapy water or spray the formula on the appliance. Rinse.

94. THE REFRIGERATOR TRAY
- Besides the basement, the most common place for mold in a house is the tray at the bottom of the refrigerator. Many people are not even aware it is there. This tray catches moisture drips from

the refrigerator and accumulates high levels of mold. Mold is then blown into the house by the appliance's fan and motor. Pulling the tray out once a month and cleaning with a borax solution is highly recommended.

95. TO CLEAN PLASTIC AND OIL COATINGS ON NEW APPLIANCES

1 teaspoon washing soda, 2 cups hot tap water

• *Step one:* Dissolve the washing soda in hot tap water either in a bowl or spray bottle. Spray or wipe the formula onto the appliance. Leave the cleaner on for an hour or so before rinsing thoroughly.

Step two: When the appliance is heated during normal use the rest of the coating will "burn off." Burn the remaining oils off in a well-ventilated area, preferably a garage. Try not to do this inside your home without ventilation or until all the new smells are gone.

Chapter 9

METAL CLEANING

Natural cleaning ingredients work like magic on metal. I used to think that one *had* to use a commercial metal polish because nothing else worked as well. One look at the gleaming results of these recipes disproved that misconception for good. Cleaning metal without poisonous chemicals is particularly important, since these items are often used for food.

Cleaning Methods

Abrasion: When a recipe calls for a granular ingredient, it is partially for its abrasive qualities. Salt and vinegar, for example, combine the grittiness of salt with the acidic qualities of vinegar. Abrasion helps rub off tarnish. It won't scratch your utensils as long as you apply the mild mineral abrasive with something soft like a sponge or rag. Don't ever use a scouring pad on metal or you will have scratches.

Acids: Commercial metal cleaners often contain strong acids. Instead of using these toxic and corrosive products, you can use naturally acidic products like lemons. Lemons work beautifully, eating up the tarnish. It is advisable, however, to give acid time to do its work. Allowing the metal to soak in an acidic solution or letting an acid-based paste of vinegar and baking soda sit on the metal for a while helps further the chemical interaction. I suggest coating the metal with the paste and going off to do something else. When you return, all you have to do is rinse the metal off with hot water and polish it dry.

Pastes: Pastes are often used for their alkaline qualities. One approach to cleaning with pastes is to cut up an old sponge and use it to rub the paste onto the metal. Or use a clean rag. For hard-to-reach areas try scrubbing pastes on with a soft toothbrush. Once the metal has been rubbed with the paste, leave it on to dry. Rinse the metal with hot water and polish dry with a soft, clean cloth.

Packaging suggestions for pastes: Make recipes in bulk and store in wide mouth glass jars. To keep stored pastes moist, cut a small sponge cube, saturate with water, and place inside the jar. Close tightly.

Submersion: Submersion is one of the easiest and most successful methods of cleaning some metals. It simply involves covering the item to be cleaned with a specially formulated solution and either simmering it on the stove or letting it sit for a while.

Aluminum

□ NATURAL INGREDIENT CHOICES
Supermarket: lemons, vinegar, cream of tartar, cornstarch, rhubarb, tomatoes
Health Food Store/Mail Order: vegetable-oil-based liquid soap
Garden: rhubarb, tomatoes
□ CLEAN & GREEN COMMERCIAL PRODUCTS
Health Food Store/Mail Order

Bon Ami
   Cleaning Powder
Dr. Bronner's
   Pure Castile Soaps
Ecover
   Cream Cleaner
   Dishwashing Liquid

Granny's Old Fashioned
   Products
   Aloe Care
Infinity Herbal Products
   Heavenly Horsetail All
   Purpose Cleaner

□ ECO BUY: Discarded lemon rinds from salad dressing or garden, wild rhubarb, or fallen tomatoes from your garden are free cleaning ingredients.

INGREDIENT HIGHLIGHTS
Air Freshener: When using cleaners which contain citrus, such as those in this chapter, stew the discarded rinds for an air freshener.

CLEANING TIPS
Time Saver: Aluminum tends to develop a greyish discoloration. If you choose any of the liquid submersion recipes, not only will they take the stain ompletely away but they will save time and effort because you don't have to do any scrubbing. It couldn't be easier.

Warning: Baking soda and washing soda should not be used on aluminum.

96. CITRUS CLEANER
2 or 3 halved lemons or 1 grapefruit cut four ways, water
• Put the cut lemons or grapefruit along with water in the tarnished pan, or put the tarnished utensil in a pan with water and fruit. Make sure that the stained sections are submerged. Stew on a low heat for an hour or so, until stains are gone. This recipe will freshen the air, too.

97. VINEGAR AND CREAM OF TARTAR
2 tablespoons cream of tartar, enough vinegar to make a paste
• Add the vinegar to the cream of tartar, drop by drop, until you have a paste. If it gets too liquid add more cream of tartar. This should be a fairly stiff paste. Rub the paste on the metal with a sponge or cloth. Let the paste dry completely before washing it off with hot water. Rub dry with a clean cloth. This method is useful for items that you cannot submerge and heat.

98. ALUM AND CORNSTARCH
1 tablespoon alum, 1 tablespoon cornstarch, enough water to make a paste
• Mix ingredients together in a bowl. Rub the paste onto the metal using a damp sponge. Let dry. Rinse in hot water and polish dry with a clean rag.

99. SPRING CLEANING WITH RHUBARB
1 to 2 cups freshly cubed rhubarb, water
• Put the rhubarb in the stained pot with enough water to cover the stains. If you are cleaning a utensil, submerge it in a pot with the rhubarb and enough water to cover. Stew until the stains are gone. Discard the cooked rhubarb.

100. HARVEST OF TOMATOES
2 sliced or halved tomatoes, water
• Follow the directions for cleaning aluminum with citrus or rhubarb. Canned tomatoes may also be used. Discard the tomatoes.

Brass, Bronze and Copper

☐ NATURAL INGREDIENT CHOICES
Supermarket: salt, flour, vinegar, baking soda, lemons, cream of tartar, buttermilk, sour milk, tomato juice, Worcestershire sauce, soap, olive oil
Health Food Store/Mail Order: vegetable-oil-based liquid soap

---

101. NEW UTENSIL LACQUER ELIMINATOR
   pan of boiling water, 2 tablespoons baking soda, 2 tablespoons washing soda
   • Submerge object in a non-aluminum pot of boiling water and minerals until the lacquer peels off.

102. BRASS AND COPPER CLEANER I
   2 teaspoons salt, 1 tablespoon flour, enough vinegar to make a paste
   • Put the salt and the flour in a small bowl or cup. Use more salt for a grainier, more abrasive paste, and less for a softer paste. Add enough vinegar to produce a thick paste texture that you like. This paste actually feels "professionally" made. Rub the paste on energetically with a clean sponge. Let dry completely, and then rinse in hot water. Polish to a high luster with a clean, soft cloth.

103. BRASS AND COPPER CLEANER II
   supermarket brand of white toothpaste
   • If you have any discolored spots left after using the formula in part I, you can eliminate them by rubbing with toothpaste. I recommend using your bare hands for this, because your body oils combine nicely with the toothpaste. Make sure your hands are dry. If you don't want to use your bare hands, add a drop or two of olive oil to the toothpaste.

104. SALT, LEMON, VINEGAR
   2 tablespoons salt, 1 tablespoon lemon juice, 1 tablespoon vinegar
   • Follow directions for Brass and Copper Cleaner I.

105. LEMON JUICE
   juice of one lemon, or more, as needed

• Rub lemon juice on with a soft cloth or old sponge. Rinse and polish dry.

106. LEMON AND SALT
1 tablespoon lemon juice, 1 tablespoon salt
• Make a paste and rub on the metal with a sponge. Rinse in hot water and polish dry with a soft dry cloth.
Alternative: Cut a lemon in half and rub the cut side into salt until the salt has stuck. Rub the lemon onto the metal. Rinse in hot water and polish dry.

107. VINEGAR AND LEMON JUICE
2 tablespoons vinegar, 2 tablespoons lemon juice
• Saturate a sponge with this mixture and rub energetically onto the metal. If the metal is really dirty, soak it in vinegar and lemon juice overnight. Rinse in hot water and buff dry.

108. VINEGAR AND SALT
3 tablespoons salt, just enough vinegar to make a paste
• Use the salt liberally to make a paste. The salt is used as an abrasive.

109. HOT MILK
equal parts milk and water
• Put enough milk and water in a pan to cover the tarnished metal. If there is a lot of tarnish, use a stronger milk ratio. Simmer gently over low heat for an hour or so. For small areas, dab milk on a cloth or sponge and rub the metal.

110. TOMATO JUICE
1 cup tomato juice
• Follow directions for Hot Milk, above.

111. CATSUP CLEANER
catsup
• Dab a generous amount of catsup onto a used but clean sponge. Rub the metal energetically with the sponge. Let it sit for an hour or so if it is tarnished badly. Rinse with hot water and soap. Rub dry.

112. WORCESTERSHIRE SAUCE CLEANER

Worcestershire sauce
• Follow directions for Catsup Cleaner.

113. CREAM OF TARTAR PASTE

2 tablespoons cream of tartar, enough water to make a paste
• Place cream of tartar in a bowl and add enough water to make a paste. Rub onto metal, let dry, rinse in hot water, and polish dry.

114. HEAVY-DUTY CLEANING

¼ cup lemon juice, ⅛ cup salt, enough water to cover metal
• Soak the tarnished piece in a solution of lemon juice, water and salt overnight. Rinse with hot water. Dry with a soft cloth.

115. SOAP AND VINEGAR

1 tablespoon vegetable-oil-based liquid soap, 1 tablespoon vinegar
• Use a recommended vegetable-oil-based liquid soap or make your own by peeling soap shavings from a bar and heating it with a little water on the stove. Apply this liquid soap and vinegar to the metal with a sponge or clean rag. Rinse. This is a good cleaner for metal that has grease on it.

116. BRASS SHINER

1 tablespoon vegetable or nut oil
• This recipe is for shining a piece you've already cleaned. Dab oil on a soft rag. Rub and buff the metal with the cloth. This is not recommended for eating utensils.

117. FOR TOUGH JOBS

• Soak item overnight in any above solution of your choice.

Chrome

□ **NATURAL INGREDIENT CHOICES**
Supermarket: baking soda, lemon, vinegar, aluminum foil

Bon Ami
  Cleaning Powder

118. CHROME CLEANER
3 tablespoons baking soda, enough water for a paste
- The abrasive cleaning power of baking soda on spots and stains is very effective. Once the chrome is cleaned, rinse well with warm water and polish.

119. LEMON RIND CLEANER
rind of one lemon
- Slice a lemon in half and make some lemonade. After the juice is used, rub the chrome with the inner white part of the lemon. Rinse.

120. CIDER VINEGAR ON CHROME
⅛ cup cider vinegar
- Saturate a sponge with cider vinegar and rub on metal. Rinse.

Gold

☐ **NATURAL INGREDIENT CHOICES**
Supermarket: baking soda, soap, toothpaste

121. GLEAMING GOLD
white supermarket-type toothpaste
- Put a ¼-inch squirt of toothpaste on your hand. Your natural body oils are part of the formula, so make sure your hand is dry. Rub the toothpaste on the gold until it is clean. Rinse well in hot water. Polish to a high luster. For filigree, use a toothbrush with toothpaste on it to get at hard-to-reach areas.
Alternative: Mix toothpaste with enough water to make a thick liquid. Saturate a sponge with the mixture and wash the gold.

122. BAKING SODA
2 tablespoons baking soda, enough water for a paste
- Rub this paste on the gold with a sponge or cloth. Rinse and polish dry with a soft cloth.

123. GOLD JEWELRY

glass jar, 3 small squirts toothpaste, ¼ teaspoon liquid soap, ¼ teaspoon baking soda, warm water

• Put the jewelry in a glass jar with soap, a few squirts of toothpaste and enough warm water to cover. Shake well. For badly tarnished gold let the concoction sit for a couple of hours. Shake some more. Rinse well, dry, and polish with a soft cloth.

Iron

□ **NATURAL INGREDIENT CHOICES**
Supermarket: alum, vinegar, vegetable oil, rhubarb
Mother Nature: hay, rhubarb
Mail Order: cork

CLEANING TIP

Protect iron utensils from rusting by keeping them dry. After washing, set them in a warm oven or over a low flame. A thin layer of vegetable oil acts as a further protectant.

124. ALUM AND VINEGAR

3 tablespoons alum, just enough vinegar to make a paste

• Put the alum in a bowl. I suggest adding the vinegar drop by drop, because you want this to be a thick paste. Rub the mixture on the metal using a mild abrasive pad such as a supermarket green pad. Rinse and dry completely. This recipe calls for a lot of alum because the paste will get discolored with rust very quickly and you will want to renew it.

125. HAY DAY

hay, ¼ cup vinegar

• Fill the rusty pot with hay, cover with water and vinegar, and boil until the rust is all gone. Repeat this procedure a couple of times for exceptionally rusty utensils. This is a truly old-fashioned recipe.

126. OILED CLOTH

cloth, 1 to 2 tablespoons vegetable oil

• Dip the cloth in the oil and rub energetically on the rust spots. You can also use a sponge or a cork.

127. RHUBARB STEW TO CLEAN A RUSTY POT

2 to 3 cups sliced rhubarb, enough water to cover the rust

• Place the rhubarb and water in a rusty pot. Stew over low heat, stirring occasionally. Once the rust lifts off, throw away the rhubarb, wash the pot, and dry thoroughly.

Pewter

☐ **NATURAL INGREDIENT CHOICES**
Supermarket: soap, cabbage, vegetable or nut oils, vinegar, flour, salt
Health Food Store: vegetable-oil-based liquid soaps
Mail Order: rottenstone, vegetable-oil-based liquid soaps
Garden: cabbage
Liquor Store: brandy, vodka
Hardware Store: soft blackboard chalk
☐ **CLEAN & GREEN COMMERCIAL PRODUCTS**
Health Food Store/Mail Order

Ecover	Wood Finishing Supply
Cream Cleaner	Company
	Rottenstone

Any Vegetable-oil-based liquid soap in Chapter 2.

INGREDIENT HIGHLIGHTS
Chalk: To make powdered chalk from chalk sticks, place the sticks in a paper bag, roll the bag up, and with a hammer pulverize the chalk to a powder. Pour the powder into a bowl and grind up any remaining lumps with a mortar and pestle or rock.

128. SALT, FLOUR, AND VINEGAR PASTE

2 teaspoons salt, 1 tablespoon flour, enough vinegar to make a paste

• Mix ingredients together in a bowl. Cover metal with the paste and let it dry. Rub off what you can with a dry cloth (for extra abrasion) before rinsing in hot water and polishing dry.

129. PLAIN OLD SOAP AND WATER

bar of pure soap, sponge

• Rub the bar of soap onto a damp sponge. Once you have

enough soap on the sponge, wash the pewter with it, rinse in hot water, and polish dry with a soft cloth.

130. THE CABBAGE PATCH
wet cabbage leaves, salt
• Cut a cabbage in half or in wedges, dampen it under the faucet, sprinkle with salt,and rub the freshly cut side onto the metal until the pewter is shiny and clean.

131. CHALK AND VODKA OR BRANDY
¼ cup ground-up soft chalk, brandy or vodka
• Put the chalk in a bowl and add enough alcohol to make a paste. Follow directions for pastes at the beginning of this chapter.

132. THE OLD ROTTENSTONE
⅛ cup fine rottenstone, enough vegetable or nut oil to make a paste, a few drops of vinegar
• Rub this paste on the metal gently, using your fingers to benefit from the full abrasive affect of the rottenstone. Keep working the paste into the metal until it is clean.

Pots and Pans

□ **NATURAL INGREDIENT CHOICES**
Supermarket: salt, vinegar, baking soda, washing soda, tomatoes, tomato juice
Garden: tomatoes
□ **CLEAN & GREEN COMMERCIAL PRODUCTS**

<u>Health Food Store/ Mail Order</u>

Bon Ami	Ecover
Cleaning Powder	Cream Cleaner

<u>Supermarket</u>

Bon Ami
Cleaning Powder
□ **ECO-BUY:** Most of these recipes either are free or cost pennies.

INGREDIENT HIGHLIGHTS
Chalk: For more about chalk, see Cleaning Tips in Chapter 3.

CLEANING TIPS

For more cleaning suggestions, determine what metal your pots are made of and look up that metal in the Table of Contents. Also, see Chapter 6, The Kitchen, for information on washing dishes. *Warning:* Baking soda and washing soda are not recommended for cleaning aluminum.

133. SALT FROM THE SHAKER
   salt, water
   • Soak in a concentrated solution of salt and warm water.

134. WASHING SODA SOAK FOR BURNT ON FOOD
   2 tablespoons washing soda, a squirt of vegetable-oil-based liquid soap, dishpan of hot water
   • Place the washing soda and soap in a dishpan. Dissolve with the hot water and blend. Soak the dirty pans in the solution until the food lifts off easily. Wash as usual.

135. BAKING SODA PASTE
   3 tablespoons baking soda, water
   • Make the paste in a small bowl. Scoop the mixture onto a sponge and rub the surface with it. Leave the paste on the metal until dry. Rinse well in hot water.

136. ENAMEL PAN CLEANER
   2 teaspoons washing soda, 2 teaspoons baking soda, water
   • Simmer ingredients and water in the pan until clean. Make sure you put in enough water to cover the stained areas. Rinse.

Silver

□ **NATURAL INGREDIENT CHOICES**
Supermarket: salt, baking soda, aluminum foil, cream of tartar, toothpaste, olive oil, milk, vinegar, lemon, lime
Health Food Store/Mail Order: vegetable-oil-based liquid soap, diatomaceous earth
Liquor Store: vodka

Ingredient Note: When a recipe calls for diatomaceous earth it does not mean the kind available in pool supply stores. For more details turn to the section on Ingredients in Chapter 22.

CLEANING TIPS

Silver Protector: To keep silver from tarnishing, store silver that is not being used out of the light, either wrapped in soft felt or cloth.

Time Saver: If you use the Magical Tarnish Remover you can do something else while the metal is cleaning itself.

Warning: Don't soak glued silver pieces.

137. MAGICAL TARNISH REMOVER

1 tablespoon salt, 1 tablespoon baking soda, a few sheets of aluminum foil

• Fill a pan big enough to hold the silver with water (a corn steamer, for example). Add the silver, baking soda, salt, and aluminum foil. Let the mixture sit for an hour or so. If you stand over it, you will notice a slightly rotten egg smell. The tarnish will magically be pulled from the silver onto the aluminum foil. Rinse the silver in hot water and polish dry with a soft cloth.

138. MAGICAL SILVER POLISH VARIATION

2 tablespoons cream of tartar, a few sheets of aluminum foil, water

• Follow directions for Magical Tarnish Remover.

139. TO MAKE PERFECT: TECHNIQUE #1

white supermarket-type toothpaste

• If you have remaining spots and tough-to-clean areas, dab some toothpaste into your hand. Your natural body oils are part of this formula, so make sure your hands are dry and haven't been washed too recently. Massage the toothpaste onto the spots. Rub it in until the discoloration is gone. The toothpaste will turn grey and start smelling like metal. A small amount of toothpaste goes a long way when mixed with oils; if you don't want to use your bare hands, add a drop or two of olive oil to the toothpaste. Toothpaste polishes silver perfectly.

140. TO MAKE PERFECT: TECHNIQUE #2

white supermarke-type toothpaste, water

• A little bit of toothpaste goes a long way with this recipe. Put a dab or so of toothpaste into a bowl and add enough water to make a paste. Rub the paste on the silver with a damp sponge until the silver is clean. Rinse in hot water and buff dry with a soft cloth.

141. BAKING SODA
baking soda, wet sponge
• I used this polish for years until I learned about the magical cleaners and toothpaste. It works very well. Make a paste of baking soda and water. Use an old sponge to rub the paste on the silver, then rinse in hot water and dry with a soft cloth.

142. OIL POLISH
vegetable or nut oil
• After cleaning the silver, polish it with oil. Dab some oil on a soft cloth and buff the silver. This isn't recommended for eating utensils.

143. MILK AND VINEGAR
equal parts milk and vinegar
• Put silver in a pan and cover with the milk and vinegar. Soak overnight. Rinse in soap and hot water and polish dry.

144. MILK AND CREAM OF TARTAR
1 tablespoon cream of tartar, 1 tablespoon salt, enough milk to cover tarnish
• Place silver and ingredients to cover in a pan. Soak overnight. Rinse in soap and hot water and polish dry.

145. LEMON AID FOR JEWELRY
juice of 1 or 2 lemons
• Soak the silver in the lemon juice. Rinse well with hot water and polish dry.

146. LIME AID FOR JEWELRY
juice of 1 or 2 limes
• Soak the silver in the lime juice. Rinse well with hot water and polish dry.

147. DIATOMACEOUS EARTH
½ cup diatomaceous earth, water
• The reason some commercial silver cleaners are greyish is that they contain diatomaceous earth. Make a paste by adding water to the diatomaceous earth. Rub the mixture gently onto the silver. Rinse with hot water and polish dry with a soft cloth.

148. VODKA, SOAP AND DIATOMACEOUS EARTH

½ cup diatomaceous earth, 3 tablespoons vodka, ½ teaspoon vegetable-oil-based liquid soap, water

• Place the diatomaceous earth, vodka and soap in a jar. Add enough water to make a thick paste. Proceed as if you were using a commercial paste. Store in a wide mouth jar with a damp sponge.

Stainless Steel

□ **NATURAL INGREDIENT CHOICES**
Supermarket: baking soda, borax, lemon juice, salt, vinegar, toothpaste
□ **CLEAN & GREEN COMMERCIAL PRODUCTS**
Supermarket
  Bon Ami
    Cleaning Cake
    Cleaning Powder

---

149. BAKING SODA ABRASION

baking soda, wet sponge, water

• Put the baking soda in a small dish. Scoop it onto a damp sponge. Rub the baking soda on the metal, continually working it until the stainless steel is clean. Rinse well in warm water and polish dry.

150. BORAX

3 tablespoons borax, enough water for a paste

• Rub this paste on as you would any other. Rinse well with warm water and polish dry with a soft cloth.

151. LEMON JUICE AND SALT

2 tablespoons salt, enough lemon juice to make a gritty paste
• Follow directions for pastes.

152. VINEGAR

vinegar, water

• Saturate a sponge with vinegar. Rub the metal until it is clean, rinse well, and polish dry with a soft cloth.

153. TOOTHPASTE: A LITTLE DAB'L DO YA
white supermarket-type toothpaste
• Follow recipe directions earlier in this chapter for silver,
To Make Perfect: Technique #1 or Technique #2.

All-Purpose Metal Cleaners

□ **NATURAL INGREDIENT CHOICES**
Supermarket: baking soda
Hardware Store: blackboard chalk
Fireplace: wood ashes
Mail Order: diatomaceous earth
□ **CLEAN & GREEN COMMERCIAL PRODUCTS**
<u>Health Food Store/Mail Order</u>

Bon Ami   Ecover
  Cleaning Powder   Cream Cleaner
<u>Supermarket</u>

Bon Ami
  Cleaning Powder
□ **ECO-BUY:** Use fireplace ashes for zero cost.

154. CHALK
equal parts soft blackboard chalk and baking soda
• Place 3 or 4 sticks of blackboard chalk in a paper bag. Roll up
the bag and pound the chalk into a powder with a hammer. Pour
the powder into a bowl and grind up any remaining lumps with
a pestle or rock. Mix with baking soda, add enough water to make
paste, and rub gently onto the metal.

155. ASHES
¼ cup powdered wood ashes
• Rub the ashes onto the metal with a damp sponge.

156. DIATOMACEOUS EARTH
3 tablespoons diatomaceous earth, 3 tablespoons baking
soda, enough lemon juice to make a paste
• Mix diatomaceous earth with baking soda and lemon juice

and rub it gently onto the metal. Rinse well with hot water. If you are unfamiliar with diatomaceous earth, see Ingredient Note for silver earlier in this chapter.

157. TOOTHPASTE: A LITTLE DAB'L DO YA
white supermarket-type toothpaste
• Follow recipe directions earlier in this chapter for silver, To Make Perfect: Technique #1 or Technique #2.

158. TABASCO SAUCE
tabasco sauce
• Shake some tabasco directly onto the metal, rub with cloth or sponge, rinse, and polish dry with a soft cloth.

Chapter 10

THE BATHROOM

Tub and Tile Cleaners

☐ **NATURAL INGREDIENT CHOICES**
Supermarket: washing soda, baking soda, vinegar, borax, cream of tartar, lemons
Health Food Store/Mail Order: vegetable-oil-based liquid soap
☐ **CLEAN & GREEN COMMERCIAL PRODUCTS**

Health Food Store/Mail Order

A.F.M. Enterprises
   X158 Mildew Control
   Safety Clean
   Super Clean
Dr. Bronner's
   Pure Castile Soap
Bon Ami
   Cleaning Powder
   Cleaning Cake

Ecover
   Toilet Cleaner
   Cream Cleaner
Greenspan
   Friendly Cleaner
Naturally Yours
   Basin, Tub, and Tile
   Cleaner
   Toilet Bowl Cleaner

Supermarket

Arm and Hammer
   Baking Soda
   Super Washing Soda
Bon Ami
   Cleaning Powder
   Cleaning Cake

Dial Corporation
   20 Mule Team Borax

INGREDIENT HIGHLIGHTS
Mineral Magic: Custom-make your own cleaners by mixing the following minerals. The bigger the job, the more mineral you need, and the more you use the more you need to rinse.
   Baking soda: odor absorbing, deodorizing
   Borax: disinfects, deodorizes, inhibits mold
   Washing soda: cuts grease

Zeolite: ion exchanger, absorbs pollutants
Pumice: stain remover, polisher

CLEANING TIPS

Cleaning fiberglass: Substitute baking soda or borax for recipes that call for washing soda. Washing soda can scratch fiberglass. Two commercial products that clean fiberglass are *Ecover Cream Cleaner* and *Bon Ami Cleaning Cake.*

Cleaning tip for hard-water areas: Vinegar or lemon juice help dissolve mineral buildup. If you really need a solvent, try *Plant Thinner*, made by Auro Natural Plant Chemistry.

Health note: Although the chemicals DEA, MEA, and TEA (among others) are suspected of causing the formation of carcinogenic nitrosamines in cosmetics, it is unclear if they do this in soaps. To be on the safe side, I suggest you add a couple of drops of liquid vitamin E (available at your health-food store) per ½ gallon to all commercial all-purpose liquid soaps and dish soaps recommended in this book to help protect against possible nitrosamine contamination. To make this task easy, just add the vitamin E to the bottle when you open it the first time. You do not have to do this at all if either vitamin E or vitamin C (ascorbic acid) is listed in the ingredients.

Note: These recipes are formulated using the vegetable-oil-based liquid soaps recommended in Chapter 2, All-Purpose Cleaners.

159. WALTER'S TUB-AND-TILE SPRAY CLEANER

½ teaspoon washing soda, 1 teaspoon borax, ¼ to ½ teaspoon vegetable-oil-based liquid soap, 3 tablespoons vinegar, 2 cups very hot tap water, spray bottle

• My friend Walter wanted me to develop a nontoxic tub cleaner that had super power, and here it is. I stood in the bathtub one winter, barefoot, spraying different formulas on the walls, until I came up with this solution. Place all the ingredients in a spray bottle and shake well. Spray the cleaner onto the tub or tile and rinse or wipe off with a damp sponge. If your spray bottle clogs up, next time use less washing soda and borax.

160. CLEANER WITH CITRUS-SCENT FRESHENER

• Choose any recipe that suits you, adding ¼ cup lemon juice or a few drops of an essential oil derived from lemon, grapefruit, or orange peel. You can make this yourself by following the directions for making essential oils in Chapter 1, Ecological Cleaning. Or, if you would prefer something other than citrus, choose your own fragrance from the large variety of fragrant herbs and essential oils available. Your choice of perfumes may change from season to season and mood to mood. If you don't have any essential oils, fragrant herbs or citrus, another option is to add a few drops of one of Dr. Bronner's naturally scented castile soaps, such as lavender, to your cleaner.

161. MILDEW AND MOLD INHIBITOR WITH DISINFECTANT
1 teaspoon borax, 3 tablespoons vinegar, 2 cups hot tap water, spray bottle
• Place the borax and vinegar in a spray bottle. Dissolve the borax by pouring the hot tap water over it. Shake to mix. Spray this formula onto mold-growing areas of your bathroom and leave on without rinsing. The vinegar will evaporate.

162. SCOURING PASTE
⅓ cup baking soda, ⅓ cup borax, 1 teaspoon vegetable-oil-based liquid soap, water to make a thick paste
• Place baking soda, borax and soap in a bowl. Add water, bit by bit, until the powder is moist. Scoop some of the paste onto a sponge or cellulose cloth and go to work. Thorough rinsing is required.

163. HEAVY DUTY TILE CLEANER
1 cup washing soda
• Scoop some washing soda onto a damp sponge, wash tiles, and rinse well. Washing soda has good grease-cutting ability and is an odor absorber. This recipe is recommended only for heavy cleaning jobs because the washing soda requires a lot of rinsing.

164. PORCELAIN CLEANER
baking soda
• Scoop the baking soda onto a sponge or cellulose cloth and rub onto the procelain until clean. Rinse well.

165. FIBERGLASS CLEANER I
½ cup borax, enough vinegar to moisten the borax
• Place borax in a bowl. Add vinegar slowly until borax is damp. Scoop onto a sponge and rub on the fiberglass. Continue rubbing until clean. Rinse thoroughly.

166. FIBERGLASS CLEANER II
½ cup baking soda, water
• Follow directions for Fiberglass Cleaner I.

Toilet Bowl Cleaners

Before embarking on nontoxic cleaning of the toilet, make sure that you have removed all long-lasting synthetic cleaning products that might be attached to the inside of the toilet tank.

□ **NATURAL INGREDIENT CHOICES**
Supermarket: borax, vinegar, baking soda, cream of tartar
Health Food Store: 1000 mg. vitamin C capsules
□ **CLEAN & GREEN COMMERCIAL PRODUCTS**

Health Food Store/Mail Order	
A.F.M. Enterprises	Ecover
Safety Clean	Toilet Cleaner

Supermarket	
Arm & Hammer	Dial Corporation
Baking Soda	20 Mule Team Borax
Super Washing Soda	

INGREDIENT HIGHLIGHTS
Studies show that these substances have disinfectant properties.
Borax: disinfects, deodorizes, inhibits mold growth
Pine oil: disinfects

167. BORAX OVERNIGHT PERFECT CLEANER
1 cup borax
• I discovered the wonders of borax as a toilet cleaner one morning quite by chance. The night before, I had wearily poured some borax into the toilet bowl but had been too tired to scrub it. My daughter was a baby at the time and I was a tired mother. I just left the borax there without even flushing. The next morning, to my amazement, the dirty rings just lifted and went away. No work; it was incredible. To accomplish no-work toilet cleaning, simply pour the borax into the bowl before going to sleep at night. In the morning the stains will be effortlessly brushed away. Of course, if you insist on scrubbing .....

168. ON-THE-SPOT CLEANING: BORAX AND VINEGAR OR LEMON JUICE
1 cup borax, ¼ cup vinegar or lemon juice
• Pour the ingredients into the bowl. Let rest for a few hours, then scrub with a toilet bowl brush. Flush.

169. VITAMIN C
1 or 2 1000 mg. vitamin C capsules
• Open a capsule and drop the ingredients into the toilet before you go to sleep. Swirl a toilet bowl brush around the rings in the morning and then flush. That is all there is to it.

170. CREAM OF TARTAR
2 or 3 tablespoons cream of tartar, water
• Sprinkle some cream of tartar on the area, scrub, and rinse.

171. BAKING SODA
½ cup baking soda
• Until I discovered the special toilet-bowl cleaning abilities of borax, I always used baking soda. Pour the baking soda into the toilet and clean as you normally would.

172. BAKING SODA AND PINE OIL
½ cup baking soda, a few drops pure pine oil
• Put baking soda and pine oil, a disinfectant, in the toilet bowl and let rest for an hour or so. Proceed by cleaning the bowl as you normally would. Pine oil's benefit is that it is a disinfectant.
Note: Pine oil is a common allergen, so please test carefully before using.

173. POTPOURRI ON THE TOILET TANK
potpourri
• A bowl of potpourri placed on top of the toilet not only looks really pretty but freshens the air with its subtle fragrance.

Faucets, Fixtures, and Shower Curtains

☐ **NATURAL INGREDIENT CHOICES**
Supermarket: vinegar, borax, lemon juice, washing soda, salt
☐ **CLEAN & GREEN COMMERCIAL PRODUCTS**
<u>Health Food Store/Mail Order</u>
Ecover
    Cream Cleaner
☐ **ECO-TIP:** Instead of buying a plastic shower curtain, buy a biodegradable, cotton one. This way you protect your health by avoiding the unhealthy fumes emitted from plastic and you are kind to the earth by reducing plastic consumption.

CLEANING TIPS
Hard-water Mineral Buildup: Increase the proportions of vinegar or lemon juice if you live in a hard-water area. The acidic properties of these ingredients help eat up mineral stains.

174. FAUCETS AND FIXTURES
⅓ cup vinegar, ⅔ cup water
• Place the water and vinegar in a bowl and mix well. Saturate a sponge or rag and polish the fixtures until clean.

175. MINERAL BUILDUP
1 teaspoon alum, ¼ cup vinegar or lemon juice
• Mix the ingredients in a bowl. Saturate a rag with the mixture and rub on the mineral buildup until clean. Or, for really bad scale, lay the saturated rag directly on the problem area and let it sit there for a few hours before rinsing.

176. SHOWER-CURTAIN VINAIGRETTE
vinegar
• Rub a sponge saturated with vinegar on your shower curtain to remove the soap buildup. Vinegar will also help kill mold and mildew.

177. SHOWER-CURTAIN MOLD REMOVER
⅓ cup vinegar or lemon juice, ⅓ cup borax
• Combine ingredients to make a paste and scrub on shower curtain using a sponge. Rinse well. To make this a little easier, just

put the curtain into the bathtub or have someone hold the bottom of the curtain taut as you scrub up and down from above.

178. SHOWER-CURTAIN CLEANER
½ cup borax, warm water
• Wash curtain in the washing machine with borax on a warm-water setting.

179. SHOWER-NOZZLE CLEANER
⅓ cup vinegar, ⅓ cup hot tap water
• Mix the water and vinegar and saturate a sponge or old tooth brush with it. Scrub directly onto the shower head.

180. MILDEW INHIBITOR (FOR SHOWER CURTAINS)
warm water, ½ cup salt, ½ cup borax
• Fill tub with warm water and minerals. Soak shower curtain for an hour or so to help prevent mildew. Don't rinse.

181. GETTING RID OF THE NEW PLASTIC SHOWER CURTAIN SMELL
sunlight
• Lay the shower curtain in the sun, turning every few hours. The plastic will out-gas and the worst of the smell will be gone in a day.

Chapter 11

FURNITURE

Wood Furniture

☐ **NATURAL INGREDIENT CHOICES**
Supermarket: lemons, lemon juice, vinegar, olive oil, vegetable oil, tea, nuts, salt, washing soda, baking soda, borax, toothpaste
Health Food Store: walnut oil, almond oil, food-grade linseed oil, vegetable-oil-based liquid soap, aromatic herbs
Mail Order: washing soda, pumice, pure lac flakes, beeswax, turpentine alternative, alcohol alternative, carnauba wax, rottenstone
Pharmacy: iodine
Hardware Store: crayons
Liquor Store: whiskey, wine
Art Supply Store: beeswax, crayons
Fireplace: wood ashes

☐ **CLEAN & GREEN COMMERCIAL PRODUCTS**
Health Food Store/Mail Order

A.F.M. Enterprises
    All-purpose Polish and
    Wax
Auro Organics Natural
Plant Chemistry
    Boiled Herbal Linseed
    Oil
    Boiled Linseed Oil
    Larchwood Balm
    Beeswax Balm
    Plant Thinner
    Plant Alcohol Thinner
    Beeswax Care
    Arve Plant Polish
    Shellacs, paints, stains,

paint removers, etc.
Bon Ami
    Cleaning Powder
Karen's Nontoxic Products
    Lemon oil
Livos
    Kiros-Alcohol Thinner
    Landis-Shellac
    Trebo-Shellac
Naturally Yours
    Furniture Cleaner and
    Protector
    Natural Solvent Spotter

Wood Finishing Supply
 Company
    Carnauba Wax
    Shellac (lac flakes)
    Beeswax
    Rottenstone
    Pumice

□ **IMPLEMENT:** Professional polishers polish with cheesecloth folded many times over upon itself.

INGREDIENT HIGHLIGHTS

Alcohol Alternatives: To avoid using hydrocarbon-derived rubbing alcohol, try vodka or *Alcohol Thinner* (as distinct from *Plant Thinner*), from Auro Natural Plant Chemistry or Livos.

Lanolin: This is the oil found in sheep wool; it is available at pharmacies.

Lemon Oil: An essential oil commonly used in commercial furniture polishes, lemon oil is easy to make or buy. Essential oils are very strong and should be used sparingly. Health food stores increasingly carry essential oils but be careful to avoid any that contain synthetic ingredients. A pure lemon oil is available from Karen's Nontoxic Products (see Mail Order Suppliers, Chapter 22). If you are in a hurry, go to your grocery store and buy a bottle of pure (not artificial) food-grade lemon extract. It will be next to the vanilla in the herb section and, like vanilla, will contain alcohol. There isn't much actual lemon oil in this product but it will do in a pinch. To make your own lemon oil, see directions for making essential oils in Chapter 1, Ecological Cleaning.

Linseed Oil: Although linseed oil is used in many formulas because it has a fast drying time, linseed-oil products available from hardware stores contain petroleum products to speed drying time even more. When heated, an impure oil can emit toxic gases, volatilize, and ignite. I feel strongly enough about pure oil to suggest buying it in a health food store instead of a hardware store. In health food stores it is sometimes known as the "Omega 3" oil. Petroleum-free boiled linseed oil (boiled linseed oil dries even faster) is available from Auro Organics.

Shellac: The two ingredients of shellac, or French polish, are lac flakes and alcohol. One can also use an alcohol alternative. Shellac can safely be used on antiques because it is removable. Lac flakes can be made into shellac by dissolving in alcohol; you don't need a more toxic solvent. Use the ratio below and let it sit for 24 hours. Clean the wood first, then, using folded cheesecloth, apply the lac and alcohol mixture in a ratio of 50:50. By the last coat, use an alcohol/lac mixture of 70:30. The idea is to put layer upon layer. Mr. Rios, former head of finishing at Sotheby's Restoration, says that one can spend forty hours on one piece and put on literally thousands of layers of shellac. But he also claims the polish will last for twenty-five years if done properly! Do not put a French polish on top of a waxed surface.

Turpentine Alternative: With hesitation I have included a few recipes in which the original formula required turpentine as a solvent. I have substituted *Plant Thinner* (made by *Auro Organics*) as an alternative product derived from citrus peels. The chemically sensitive should test this product carefully before using. Remember, Plant Thinner is flammable.

CLEANING TIPS

Ratios: Determining the best ratio of ingredients for your needs may take a little experimentation. For example, you can make a "heavy" oil-and-vinegar polish by using as much as three times more olive oil than vinegar, which is appropriate for very clean, well-dusted and varnished furniture. Or you can make a "lighter" polish by using up to three times more vinegar (or lemon juice) than olive oil. The vinegar brings up dirt and the olive oil enriches the wood. I make an even more vinegary version for a homemade dusting aid, using ¼ cup vinegar and ½ to 1 teaspoon olive oil. The wood has a wonderful nutty smell for a while, and then the vinegar completely evaporates, leaving the wood not only clean but beautiful. The first time I used this I was cleaning because - why else? - my mother was coming. The first thing she said was "Oh Annie! How beautifully polished everything is!" I couldn't have done better with a commercial product.

Caution: Vinegar can dissolve the pre-existing wax on furniture. If you want to retain the wax, reduce the amount of vinegar called for in the upcoming polishing and dusting formulas.

Wood Furniture Care: Dusting

182. WOOD FURNITURE DUSTING AND CLEANING CLOTH
½ teaspoon olive oil, ¼ cup vinegar or lemon juice, soft cotton rag
- Mix the ingredients in a bowl. Dab a soft rag into the solution and dust, polish, and shine your wooden furniture with it. You can reuse this rag over and over again.

183. LIGHT AND LEMONY DUSTING CLOTH
2 or 3 tablespoons lemon juice, a few drops olive or food-grade linseed oil, soft cloth
- Place the lemon juice in a bowl, add a few drops of oil, and saturate dusting cloth with the liquid. Use the cloth to dust with.

184. FRAGRANT HERB DUSTING CLOTH
- Follow directions for Dusting And Cleaning Cloth, but add a drop or two of an essential oil.

Cleaning Before Polishing

185. GREASE AND DIRT REMOVER
½ teaspoon vegetable-oil-based liquid soap, ¼ cup vinegar, warm water
- Mix ingredients in a bowl and saturate sponge. Scrub wood with saturated sponge, rinse with warm water, and proceed to a polish. Be advised that some wood finishes are not waterproof.

Polishing

186. MY FAVORITE WOOD CLEANER AND POLISH
⅛ cup food-grade linseed oil, ⅛ cup vinegar, ¼ cup lemon juice
- Mix the ingredients in a glass jar. Using a soft cloth, rub into the wood until it is clean. This is my favorite polish. The rich nutty smell of the linseed oil is balanced by the light smell of lemon juice. Add a few drops of Vitamin E, cover, and save leftovers.

187. OLIVE OIL, VINEGAR AND AROMATIC-HERB POLISH

¼ cup olive oil, ¼ cup vinegar, drop or two of lemon oil or other aromatic herb of your choice

• Put the ingredients in a small glass jar and shake well. Using a soft 100 percent cotton cloth, wipe the mixture on the furniture until you have achieved the polish you want.

188. LINSEED OIL, VINEGAR AND LEMON OIL POLISH

¼ cup food-grade linseed oil, ¼ cup vinegar, a few drops lemon oil

• Proceed as for Olive Oil, Vinegar and Aromatic Herb Polish.

189. THE COLOR OF WHISKEY

¼ cup linseed oil, ⅛ cup lemon juice, ⅛ cup whiskey

• Proceed as for Olive Oil, Vinegar and Aromatic Herb Polish.

190. LEMON JUICE AND THE VEGETABLE OIL

⅛ cup vegetable oil of your choice, ⅛ cup lemon juice

• Proceed as for Olive Oil, Vinegar and Aromatic Herb Polish.

191. WALNUT OIL AND LEMONS

⅛ cup walnut oil, ⅛ cup lemon juice

• Proceed as for Olive Oil, Vinegar and Aromatic Herb Polish.

192. LEMON OIL AND WALNUT

¼ cup walnut oil, a few drops lemon oil

• As any chemically sensitive individual will tell you, commercial lemon-oil furniture products can be very toxic. Make your own lemon oil (see Chapter 1, Ecological Cleaning) or buy a bottle of food-grade lemon oil at the grocery store. Combine the lemon oil with walnut oil in a small glass jar. Using a soft cloth, gently rub the mixture into your furniture. This is a workable alternative. Add a few drops of liquid vitamin E to the remaining polish and save it for next time.

193. LINSEED AND LEMONS

¼ cup food-grade linseed oil, lemon juice as you like

• For a heavier polish use only a small amount of lemon juice. The lemon juice will bring up dirt, so if the furniture is dirty you might use lemon juice with just a teaspoon or so of oil. Follow the principles described in Cleaning Tips, above.

194. LINSEED AND LEMON OIL

1 cup food-grade linseed oil, lemon oil for scent

• Combine the ingredients in a small glass jar. If you are using a real essential oil, remember they are very strong. Rub into furniture with a cloth until you get a shine that you like.

195. WOOD CLEANER WITH POLISH

¾ cup *Plant Thinner*, ¾ cup food-grade linseed oil

• Mix ingredients in a glass jar. Shake well. Saturate a sponge or cloth with the mixture and rub well into the wood. If you need to rinse the wood, dampen a cloth with vinegar and wipe.

196. SOAPY POLISH

1 cup food-grade linseed oil, 1 teaspoon vegetable-oil-based liquid soap, ½ cup water, oil from an aromatic herb (optional)

• Put the ingredients in a glass jar and shake well. Saturate a cloth with the solution and proceed as you normally would when cleaning and polishing wood.

197. STRAIGHTFORWARD POLISH

vegetable or nut oil like linseed, olive, walnut or almond

• Wet a cloth with oil and rub over furniture.

198. TEA FOR TWO

1 strong cup of black leaf tea

• Brew a strong cup of tea, let it cool, saturate a cloth with it, and rub it onto your furniture.

199. VINEGAR AND LEMON OIL CLEANER

vinegar, a few drops lemon oil

• Combine vinegar and lemon oil in proportions that suit you and rub into furniture with a cloth.

200. WHEN THE WOOD REALLY NEEDS CLEANING

⅓ cup food-grade linseed oil, 3 tablespoons *Plant Thinner*

• Mix the ingredients in a glass jar. Shake well. Saturate a rag with the solution and rub into the wood well, cleaning as you go.

201. FURNITURE CLEANER

2 cups cooled black leaf tea, ½ cup vinegar

• Saturate a sponge with the mixture and wash the furniture with it. Rinse if you want.

202. SHELLAC AND WINE POLISH
shellac, wine
• Follow directions above.

203. SHELLAC AND PLANT ALCOHOL THINNER
shellac, *Plant Alcohol Thinner*
• Follow directions above.

204. SHELLAC AND ALCOHOL
shellac, alcohol
• Follow directions above.

Waxing

If you are like me, you thought you had to be in a laboratory to make wax. Not only is this not true, but you can quickly learn to custom-make waxes for different situations. Once you have the required ingredients on hand, try experimenting. For example, carnauba is one of the hardest waxes there is, so the more carnauba in your recipe, the harder the resulting wax will be. If the wax is too hard though, it will be difficult to rub onto the furniture, which is one reason why commercial waxes have solvents in them. Annie's Beautiful Wood Furniture Wax makes a wax that is a little harder than a typical paste wax and rubs into furniture beautifully. If you need a hard-as-nails finish, try putting 2 tablespoons of carnauba instead of 1 tablespoon into the recipe. Linseed oil is the oil of choice (although other vegetable and nut oils are acceptable) because it dries more quickly than other oils. Beeswax gives a very rich, characteristic fragrance to the wax, but lemon oil is famed for its beneficial effects on wood, so you might want to use some of it, too.

205. ANNIE'S BEAUTIFUL WOOD FURNITURE WAX
½ cup food-grade linseed oil, 1 tablespoon carnauba wax, 1 tablespoon beeswax, ¼ cup vinegar, food-grade lemon oil or homemade essential oil of your choice
• Put all the ingredients in the top half of double boiler (set over water) on low heat. Heat the ingredients until all the waxes

are melted. Stir to blend. Pour the hot ingredients into a wide mouth heat-resistant container and let cool completely until the wax has solidified. The vinegar will sink to the bottom. Once it has solidified, either pop it out of the container (if plastic) or cut it in half and remove. Making sure that your furniture is clean first, rub the wax onto the furniture. Next take a soft rag and dip it into some vinegar (either fresh from the bottle or from the residue in the container) and rub, buff, and polish the wax to a high shine. The vinegar cuts the oil and makes the wax perfectly smooth. This wax works very well on unfinished, well-sanded wood, too.

206. CARNAUBA, LINSEED & LEMON OIL
1 to 2 tablespoons carnauba wax, 1 cup food-grade linseed oil, a few drops lemon oil
• Put all the ingredients except the lemon oil in a saucepan and place over low heat until the carnauba wax is melted. Stir to blend. Add the lemon oil and let the mixture cool. This mixture will be a little softer than vaseline, so it is a very pasty-type wax.

Wood Furniture Problem Solving

207. POLISH AND WAX REMOVER
vinegar, water
• Use a very strong vinegar solution. Saturate a sponge and rub energetically.

208. WATER STAINS
1 tablespoon corn oil, enough salt to make a paste
• Make the paste only as grainy as you feel comfortable using on your furniture. Rub it on with your finger or use a soft rag.

209. NUT OIL FOR CUP RINGS
walnut or other nut
• Break nut into large pieces and rub the freshly broken edge of nut meat on the furniture.

210. WATER STAINS, SPOTS, RINGS I
1 tablespoon vegetable oil, wood ashes to make a paste
• Rub the paste gently onto the affected area.

211. WATER STAINS, SPOTS, RINGS II
toothpaste
• Rub a little white toothpaste on the spot.

212. WATER STAINS, SPOTS, RINGS III
1 tablespoon beeswax, ½ cup olive, linseed, or almond oil
• Heat the beeswax and olive oil gently on the stove. Stir until well blended. Pour the mixture into a heat-resistant bowl. Let the wax cool, scoop it into a wide mouth container, then rub it onto the stained area. Buff with a vinegar-dabbed rag to cut the oil.

213. CRAYON HELP
appropriate colored crayon
• Cover up scratches by rubbing with a crayon.

214. SCRATCHES
food-grade linseed oil, nut or other vegetable oil
• Rub oil into scratch with tip of your finger.

215. SCRATCHES ON DARK BROWN WOOD
vinegar, iodine
• Put a few drops of vinegar in a small bowl. Add enough iodine to make an appropriate color. Rub this mixture onto scratches using a cloth or toothpick. If you don't want to use vinegar, a bit of vodka or whiskey will do.

216. SCORCH MARKS I
finest-grade steel wool, vinegar
• Rub gently with the steel wool. Wipe away the steel wool particles with a sponge or cloth saturated with vinegar.

217. SCORCH MARKS II
¼ cup rottenstone or pumice, enough pure linseed oil to make a paste
• Mix enough linseed oil with the rottenstone or pumice to make a paste and rub onto burned area.
Note: Rottenstone is finer than pumice.

218. ALCOHOL SPILLS
olive, linseed, walnut or almond oil
• Rub oil onto the stain with your finger or a soft cloth.

219. DIRTY WOOD
2 teaspoons cornstarch, 1 cup water, spray bottle
• Place the cornstarch in the spray bottle and dissolve with warm water. Spray on the wood and let dry. Wipe off. Particularly effective on white woodwork.

220. WOOD DISINFECTANT
1 teaspoon borax, ¼ teaspoon vegetable-oil-based liquid soap, 1 cup hot tap water
• Place ingredients in a spray bottle and dissolve with very hot tap water. Spray formula on the wood, rinse, and wipe dry.

221. WOOD LIGHTENER
lemon juice
• Saturate a sponge with lemon juice and wash the wood. There is no need to rinse.

Butcherblock Care

222. VINEGAR CLEANER FOR BUTCHERBLOCK
vinegar
• Pour straight vinegar onto the butcherblock. Scrub. Rinse.

223. BAKING SODA DEODORIZER FOR BUTCHERBLOCK
baking soda
• Pour baking soda onto the block. Scrub the soda into the wood with a damp sponge. Rinse well.

Leather Polishes and Cleaners

□ **NATURAL INGREDIENT CHOICES**
Supermarket: vinegar, vegetable oil, olive oil, lemons, skim milk, egg white
Health Food Store: food-grade linseed oil, vegetable-oil-based liquid soap
Mail Order: beeswax
Art Store: beeswax
Pharmacy: castor oil, lanolin, alcohol

□ CLEAN & GREEN COMMERCIAL PRODUCTS
Health Food Store/Mail Order

Auro Organics Natural
Plant Chemistry
    Leather Care Cream
    Plant Thinner
    Boiled Linseed Oil
    Boiled Herbal Linseed
    Oil

Karen's Nontoxic Products
    Lemon oil
Wood Finishing Supply
Company
    Carnauba Wax
    Beeswax

---

224. LINSEED LEATHER POLISH
½ cup food-grade linseed oil, ½ cup vinegar, few drops liquid vitamin E
• Put the ingredients in a glass jar and shake. Saturate the end of a soft cloth and rub it onto the leather. Make sure to test on an inconspicuous area first.

225. LEATHER POLISH
½ teaspoon vegetable-oil-based liquid soap, ½ cup food-grade linseed oil
• Mix ingredients in a glass jar and proceed as with Linseed Leather Polish.

226. BEESWAX AND OIL POLISH
2 tablespoons beeswax, 1 cup food-grade linseed oil
• Combine the ingredients in a saucepan and heat slowly until the beeswax has melted. Pour the mixture into a heat-resistant bowl and let cool. Once the wax has solidified, it will be like thin vaseline. Rub the wax onto the leather, polishing with a soft 100 percent cotton cloth as you go.

227. LEMON AND OLIVE POLISH
¼ cup olive oil, a few drops of lemon oil
• Put the ingredients in a glass jar and shake well. Saturate a cloth with the mixture and buff and polish the leather, making sure to rub the oil in well as you go.

228. YOUR OWN SADDLE SOAP
⅛ cup vegetable-oil-based liquid soap, 4 tablespoons beeswax, ⅛ cup food-grade linseed oil, ¼ cup vinegar
• Heat the beeswax and vinegar in a saucepan until the

beeswax is melted. In a separate bowl combine the soap and linseed oil and add it to the saucepan. The beeswax will solidify a bit as these cooler ingredients are added, but as soon as the beeswax is liquid again, pour the entire mixture into a heat-resistant bowl. Let the mixture cool until solid. Rub the solid wax onto the leather, polishing as you go with a soft cloth.

229. LEATHER POLISH
¾ cup vinegar, ¼ cup food-grade linseed oil
• Rub mixture onto leather with a soft cloth.

230. CASTOR OIL FOR LEATHER
¼ cup castor oil
• Pour castor oil onto a cotton cloth and rub into leather, working in the oil as you go along.

231. OMEGA 3 POLISH
½ cup linseed oil, ¼ cup white vinegar, ¼ cup water
• Put all the ingredients in a glass jar and shake well. Saturate a soft cloth with the mixture and rub the cloth onto the leather, polishing as you go.

232. ALCOHOL AND VINEGAR LEATHER CLEANER
1 tablespoon vinegar, 1 tablespoon alcohol, ¼ teaspoon vegetable oil, ¼ teaspoon vegetable-oil-based liquid soap
• Mix the ingredients together in a jar or bowl. Saturate a clean sponge or cloth with the mixture and wash the leather. Try this on dirty leather shoes!

233. LEATHER POLISH
¼ cup lanolin, ¼ cup food-grade linseed oil
• Mix the lanolin and oil in a bowl and blend well. Rub onto leather with a soft, cotton rag.

234. EGG WHITE CLEANER
1 beaten egg white
• Rub the beaten egg white into the stain with a clean cloth. Rinse well.

235. BREAD ALONE FOR SUEDE
bread
• Rub the suede with bread.

236. LEMON CLEANER
½ sliced lemon
- Holding onto the rind, rub the leather with a lemon half.

Vinyl and Plastic Upholstery Cleaners

□ NATURAL INGREDIENT CHOICES
Supermarket: baking soda, vinegar, soap flakes, washing soda
Health Food Store: pure soap for shavings, vegetable-oil-based liquid soap
Mail Order: washing soda

□ CLEAN & GREEN COMMERCIAL PRODUCTS
Health Food Store/Mail Order

A.F.M. Enterprises	Ecover
Vinyl Block	Cream Cleaner

---

237. TRIED AND TRUE VINYL AND PLASTIC SMELL ELIMINATOR
baking soda, water
- Put some baking soda in a bowl and add enough water to make a paste. Scoop the paste onto a sponge and rub the vinyl with it. This requires thorough rinsing.

238. TRIED AND TRUE WITH SOAP
¼ cup baking soda, 1 teaspoon vegetable-oil-based liquid soap, water
- Mix ingredients in a bowl, adding enough water to make the soap sudsy. Scoop the paste onto a sponge and wash the vinyl. Make sure to rinse well.

239. VINEGAR ON VINYL
¼ cup vinegar, ¼ teaspoon vegetable-oil-based liquid soap, ¼ cup water
- Combine the ingredients, saturate a sponge, and wash the vinyl. Rinse.

240. NEW VINYL SMELL ELIMINATOR
unperfumed soap flakes
- Washing new vinyl seems to clean the smell from the surface

for a while, but it needs to be repeated frequently to keep the smell down.

241. NEW CAR SMELL
2 or more breather bags of zeolite
• Put zeolite in a new car to help absorb the smell of new materials.

242. CHALK IT UP TO BAKING SODA
½ cup baking soda, ¼ cup ground-up soft chalk (see recipe 161. Chalk), enough vinegar to make a paste
• Scoop the paste onto a sponge and wash. Thorough rinsing is required.

243. VINYL CLEANER
washing soda, enough water to make a paste
• Scrub the vinyl with the paste and let sit for a few minutes. Only use this for tough jobs as thorough rinsing is required.

244. A.F.M.'S VINYL BLOCK
• To seal in the vinyl chemical smell, use *A.F.M. Vinyl Block.* Follow the manufacturer's directions.

Fabric Upholstery Cleaners

□ **NATURAL INGREDIENT CHOICES**
Health Food Store/Mail Order: vegetable-oil-based liquid soap

CLEANING TIPS
For specific stain removal suggestions see Chapter 17, Stains and Tough Jobs. Spot test the stain removal recipe in an inconspicuous part of the upholstery.

245. UPHOLSTERY FOAM CLEANER
¼ cup vegetable-oil-based liquid soap, 3 tablespoons water
• Whip ingredients together in a bowl with a whisk. Pull foam off the top and rub into uphostery. Rinse well.

Formica®

☐ **NATURAL INGREDIENT CHOICES**
Supermarket: salt, vinegar, club soda, baking soda
Health Food Store/Mail Order: food-grade linseed oil,
vegetable-oil-based liquid soap
☐ **CLEAN & GREEN COMMERCIAL PRODUCTS**
Health Food Store/Mail Order
Ecover
    Cream Cleaner

---

246. FORMICA® CLEANER
½ teaspoon vegetable-oil-based liquid soap, 3 tablespoons
vinegar, ½ teaspoon linseed or olive oil, ½ cup warm water
• Put all the ingredients in a spray bottle and shake well. Spray
cabinets and wash with a sponge. Rinse well.

247. SALT AND WATER
½ cup salt, enough water to make a paste
• Put salt in a bowl and add water bit by bit till you have a
thick paste. Scoop the damp salt onto a sponge and rub the
cabinets. Rinse well.

248. CLUB SODA OR SELTZER
club soda or seltzer, sponge
• Pour the soda directly onto a sponge and wash the cabinets.
Rinse.

249. COUNTER CLEANER
¼ cup baking soda, enough water for a paste
• Scrub the paste onto the stains. Rinse thoroughly.

Chapter 12

FLOORS

Floor Cleaners

□ **NATURAL INGREDIENT CHOICES**
Supermarket: club soda, vinegar, cornstarch, borax, vegetable oil, baking soda, kitty litter, cornmeal, tea leaves, washing soda
Health Food Store/Mail Order: vegetable-oil-based liquid soap, washing soda, pennyroyal, fragrant herbs, essential oils
□ **CLEAN & GREEN COMMERCIAL PRODUCTS**
<u>Health Food Store/Mail Order</u>

A.F.M. Enterprises
   Super Clean
   Safety Clean
Auro Organics
   Floor Wax-Balm Cleaner
   Floor Wax-Balm
   Plant Soap
   Cleansing Emulsion
   Plant Thinner
   Plant Alcohol Thinner
   Boiled Linseed Oil
   Boiled Herbal Linseed Oil

Bon Ami
   Cleaning Powder
Karen's Nontoxic Products
   Lemon oil
   Pine oil
Livos
   Avi-Soap Concentrate
   Kiros-Alcohol Thinner
   Latis-Natural Concentrate

See Chapter 2, All-purpose Household Cleaners, for recommended brands of vegetable-oil-based liquid soaps.
□ **IMPLEMENTS:** buckets, pails, cotton mops, cellulose and cotton sponges

---

CLEANING TIPS
Pine Oil: A few drops of pine oil will give disinfectant properties to your formula. Pine oil is highly allergenic, so test carefully.
Warning: Washing soda can eat up wax.

250. ALL-PURPOSE FLOOR CLEANER I
⅛ cup vegetable-oil-based liquid soap, ½ cup vinegar, 2 gallons warm water
• Put soap and vinegar in the bottom of a bucket. Fill the bucket with warm water, swishing the ingredients around a bit to activate the soap. Wash the floor as you normally would.

251. ALL-PURPOSE FLOOR CLEANER II
1 cup vinegar, 1 pail water
• Wash the floor as you normally would.

252. BORAX FLOOR DISINFECTANT
½ cup borax, 2 gallons hot water
• Pour the borax into the bottom of a bucket and pour the hot water over it to make sure it dissolves well. Proceed as usual.

253. PINE OIL DISINFECTANT
⅛ cup vegetable-oil-based liquid soap, a few drops pine oil, 2 gallons hot water
• Blend the ingredients in a pail or bucket. Proceed as usual.

254. GREASE-CUTTING FLOOR CLEANER
1 tablespoon vegetable-oil-based liquid soap, ¼ cup washing soda, ¼ cup vinegar, 2 gallons hot water
• Combine ingredients in a bucket or pail, making sure to dissolve the washing soda completely. Wash as usual. Not recommended for waxed floors.

255. ANNIE'S FAVORITE WOOD-FLOOR SOAP
⅛ cup vegetable-oil-based liquid soap, ¼ to ½ cup vinegar or lemon juice, ½ cup fragrant herb tea, 2 gallons warm water
• Combine ingredients in a pail or bucket. Swirl the water around until it is sudsy. Proceed as normal.

256. FLEA REPELLANT FLOOR SOAP
4 lemons, a few drops pennyroyal, 1 tablespoon vegetable-oil-based liquid soap, 1 gallon warm water
• Slice the lemons, place in a pan with just enough water to cover, and simmer for an hour. Squeeze lemons well, strain, and pour into a bucket. Add the pennyroyal, soap, and water. Mix well. Mop onto the floor. Let dry. Rinse thoroughly.
Note: Pregnant women should not include pennyroyal.

257. NO-WAX LINOLEUM FLOOR CLEANER

⅛ cup vegetable-oil-based liquid soap, 2 gallons water
• Wash as usual.

258. CORNSTARCH LINOLEUM CLEANER

½ cup cornstarch, ½ teaspoon vegetable-oil-based liquid
soap, ½ cup water
• Place the ingredients in a bowl and stir to blend. Scoop the
paste onto a sponge and shine the floor with it. Rinse well.

259. CORNSTARCH AND WATER LINOLEUM POLISH

6 tablespoons cornstarch, 1 cup water
• Blend cornstarch and water together and polish the floor.

260. WASHING SODA FLOOR CLEANER

¼ cup washing soda, 2 gallons water
• Add washing soda to water in a bucket. Proceed as usual.
Note: Washing soda can eat up wax.

261. CONCRETE SCRUBBER

⅛ to ¼ cup washing soda, 1 or 2 tablespoons
vegetable-oil-based liquid soap, 2 gallons water
• Dissolve washing soda and soap in warm water. Wash the
floor, scrubbing particularly dirty or greasy areas with a brush. For
stains, scrub a washing soda paste on straight, but be forewarned
that thorough rinsing will be required.

262. DAMPEN THE DUST

• Before sweeping, moisten the floor with water, using a spray
bottle.

263. HOLD DOWN THE DUST FOR SWEEPING

• Toss some freshly cut grass, cornmeal or damp tea leaves
around the floor before sweeping.

264. FLOOR WAX REMOVER THAT WORKS

washing soda, water, ¼ cup vinegar in the rinse water
• Cover the floor with a thick coat of washing soda and water.
Let dry completely before scrubbing it off. If you really need to get
the wax up, be tenacious and do not skimp on the washing soda.

Washing soda needs to be rinsed very well when you use this concentrated amount, but adding vinegar to the rinse water should help pick up some of the residue.

265. HEEL SKID MARKS
baking soda, water
• Make a paste and scrub onto the marks until they are gone. Rinse.

266. HEEL SKID MARKS II
eraser
• Rub the marks away.

267. GARAGE FLOOR ENGINE OIL SPILLS
nondeodorized pure kitty litter, washing soda
• Cover the oil completely with the kitty litter. Rub it in so that the oil is really absorbed. Sweep it up and cover again until the oil has gone. Whatever residue is left will be eliminated if cleaned with washing soda.

268. FLOOR CLEANER WITH FRAGRANT HERBS
fragrant herb teas, essential oils, steeped flowers, floor cleaner formula of your choice
• Add one of these ingredients to floor cleaner formula of choice. If you use an essential oil add just a few drops.

269. LINOLEUM & LEMONS
linoleum cleaner of your choice, lemon juice
• Add lemon juice to the formula. Lemon juice leaves a lovely, light smell in the air.

Floor Wax and Polishes

□ **NATURAL INGREDIENT CHOICES**
Supermarket: club soda, vinegar, vegetable oil, lemon juice
Mail Order: beeswax, carnauba wax, turpentine alternative
Art Store: beeswax

☐ CLEAN & GREEN COMMERCIAL PRODUCTS
Health Food Store/Mail Order

Auro Natural Plant
  Chemistry
    Floor Wax-Balm Cleaner
    Floor Wax-Balm
     Plant Soap
    Cleansing Emulsion
    Plant Thinner
    Plant Alcohol Thinner
    Boiled Linseed Oil
    Boiled Herbal Linseed
     Oil

Karen's Nontoxic Products
  Lemon oil
  Pine oil
Livos
  Kiros-Alcohol Thinner
  Landis-Shellac
  Trebo-Shellac
Wood Finishing Supply
 Company
  Beeswax
  Carnauba Wax

INGREDIENT HIGHLIGHTS
See Ingredient Highlights in Chapter 11, Furniture.

CLEANING TIPS
Wax Making Tips: Annie's Floor Wax is a wax that, once it dries, is about the consistency of a moderately soft alpine ski wax. It rubs onto the floor well. Adding more carnauba will make a hard-as-nails finish, but be cautious about adding too much carnauba or it will be very difficult to rub the wax onto the floor. Linseed oil is the oil of choice because it dries more quickly than other oils. Beeswax gives a very rich, characteristic fragrance to the wax, but you might want the scent of lemon oil, too.

270. ANNIE'S FLOOR WAX
1 cup food-grade linseed oil, 4 tablespoons carnauba wax, 2 tablespoons beeswax, ½ cup vinegar, a few drops food-grade lemon oil or homemade essential oil of your choice (optional)

• Put all ingredients in a saucepan or the top of a double boiler. Place over low heat, stirring occasionally to blend as it melts. Once the waxes have all melted, stir well and pour into a heat-resistant bowl. Let the wax harden. Once the wax has hardened, remove it from the bowl and rub it onto the floor. Saturate a soft rag with vinegar using either fresh vinegar or the residue from the wax at the bottom of the bowl. Rub and polish the wax into the floor using the vinegar saturated rag.

Note: You may want to skip the added fragrance. I actually prefer the rich smell of beeswax.

271. BEESWAX AND TEA
• Add some fragrant herb tea or a few drops of essential oil to the formula of your choice.

272. WOOD-FLOOR WAX
4 tablespoons beeswax, 8 tablespoons carnauba wax, 2 cups food-grade linseed oil, ¼ cup *Plant Thinner*
• Combine the beeswax, carnauba wax, and linseed oil in the top of a double boiler. A double boiler is imperative. This formula could be flammable, so never leave the stove unattended, and heat very, very slowly. When the waxes have melted, stir to blend, add the *Plant Thinner*, and as soon as the waxes have turned to liquid again, take the pan off the heat and pour into a heat-resistant bowl. Let cool. Apply as you normally would.

273. BEESWAX FLOOR POLISH
6 tablespoons beeswax, 3 cups food-grade linseed oil
• Put the ingredients in the top half of a double boiler (over water), and heat slowly until the wax is melted. Stir to blend. Pour directly into a heat-resistant container. This polish will be the consistency of vaseline.

274. CLUB SODA OR SELTZER
club soda or seltzer, mop or sponge
• Wash the floor and polish with a dry, clean cloth.

275. OLIVE, ALMOND, WALNUT OR LINSEED OIL POLISH
oil of choice
• Put oil in a small glass jar and drop some liquid vitamin E into it as a preservative. Saturate a soft cotton cloth with the oil and gently polish the wood.

276. LEMONS AND LINSEED I
1 cup food-grade linseed oil, a few drops lemon oil or 1 tablespoon lemon juice
• Put ingredients in a glass jar and shake to mix. Saturate a cloth with the mixture and polish the floor.

277. LEMONS AND LINSEED II
1 cup food-grade linseed oil, 1 cup vinegar, 1 cup lemon juice
• Follow directions for Lemons and Linseed I.

Chapter 13

CARPET CLEANING

Most commonly available carpets out-gas noxious fumes. I know of one case where after wall-to-wall carpeting was installed all the houseplants died and the entire family of four ended up in the hospital because of the toxicity of the carpet and installation materials. At least two of the family are now chemically sensitive. This is not a unique case. I cannot state strongly enough that ordinary carpeting is to be approached with caution or avoided.

Carpet shampoos can also include strong chemicals. The installer of the carpet mentioned above was very accomodating in trying alternative rug-cleaning methods. I have found that with a safe soap, steam cleaning is successful. The steam reaches 220^0 F. and kills dust mites, fleas, and other organisms that are in the carpet. If the machine used is a good, heavy-duty, industrial model and is used properly, all the dirt will be pulled out until the water runs clear. I rejected the soap and deodorizers the installer normally used; amazingly, the brand of nontoxic dishwashing soap I use not only was compatible with the industrial machine but according to the installer did a better job in cleaning than his did!

The soap we used successfully for steam cleaning carpets is *Heavenly Horsetail All Purpose Cleaner* made by Infinity Herbal Products. If you already have a toxic carpet there is a carpet sealant on the market manufactured by A.F.M. Enterprises called *Carpet Guard*, which when sprayed on the carpet seals in the toxic agents in new rugs. A.F.M. also manufacturers a low-toxic carpet adhesive and carpet shampoo.

It is also remarkable what one can do with just minerals when you are cleaning your carpet. Borax cleans mold and disinfects, baking soda absorbs odors of all sorts, and zeolite picks up chemicals, wood smoke, and pollutants. Make sure the rug is dry, then sprinkle with the substance until the rug is fairly well covered. Rub the powder into the nap and let it rest for twenty-four hours. Vacuum well.

□ NATURAL INGREDIENT CHOICES

Supermarket: club soda, cornstarch, baking soda, white vinegar, cornmeal, borax, cabbage, salt, potatoes, soda water
Health Food Store: vegetable-oil-based liquid soap, vegetable glycerin
Mail Order: washing soda, vegetable-oil-based liquid soap, alcohol or alternative
Mother Nature: snow

□ CLEAN & GREEN COMMERCIAL PRODUCTS

Health Food Store/Mail Order

A.F.M. Enterprises
  Carpet Shampoo
  Carpet Guard
  Carpet Adhesive
Auro Organics
  Plant Thinner
  Plant Alcohol Thinner

Granny's Old Fashioned
  Products
    Karpet Klean
Infinity Herbal Products
    Heavenly Horsetail All
    Purpose Cleaner
Livos
    Kiros-Alcohol Thinner

Cleaning

278. BAKING SODA ODOR ABSORBER

box of baking soda

• Sprinkle baking soda generously over stain or entire rug. Let sit overnight, then vacuum. Baking soda is especially effective against pet odors and food spills.

279. CARPET CLEANING FOAM

¼ cup vegetable-oil-based liquid soap, 3 tablespoons or more water

• Whip ingredients together in a bowl with an egg beater. Rub the foam into problem areas of the rug. Rinse well.

280. CORNMEAL AND BORAX DISINFECTANT

1 cup cornmeal, 1 cup borax, ½ cup baking soda

• Sprinkle the mixture over the rug and then rub with a cloth. Let rest for a few hours or overnight, then vacuum thoroughly.

281. CABBAGE PATCH
head of cabbage, salt
• Cut cabbage in wedges or in half, sprinkle with salt, and rub the freshly cut side against the rug. Vacuum thoroughly.

282. SALT AND VINEGAR
⅛ cup salt, ¼ cup vinegar
• Blend ingredients in a bowl. Rub on the rug with a sponge. Let dry. Vacuum.

283. RAW POTATO CLEANER
potatoes cut into large cubes
• Rub the freshly cut side of the potatoes into the rug until the rug is clean. Discard potatoes.

284. DEEP FREEZE DANCE
• In freezing weather, when fresh snow is on the ground, put your carpet outside until all the dirt freezes. Beat it with a broom or dance around on it, shake it out, turn it over, and do it again.

285. HEAVY DUTY RUG CLEANER FOR COLORFAST RUGS
¼ cup salt, ¼ cup borax, ¼ cup vinegar
• Put ingredients in a bowl and blend to a paste. Rub paste into carpet stains and leave for a few hours. Vacuum thoroughly.

286. ZEOLITE FOR ODORS
zeolite powder
• Sprinkle zeolite powder over the rug and brush into the nap. Let the zeolite rest for 24 hours before vacuuming.

287. RUG BRIGHTENER
½ cup alum, ½ gallon hot water
• Wash the mixture into the nap of the rug. Let dry. Sweep or vacuum thoroughly.

Stains

Try to clean stains when they are still wet and fresh because they will be easier to remove. For a more complete list of stain removing suggestions, refer to Chapter 17, Stains and Tough Jobs.

288. MUD ON RUGS
salt
• Rub salt on the mud. Let it rest for an hour or so, then sweep or vacuum thoroughly.

289. COFFEE-STAIN REMOVER
club soda, salt
• Rub club soda into the spot. Sprinkle with salt. Let the mixture rest for a few minutes, then rinse with a damp sponge.

290. ALL-PURPOSE SPOT REMOVER
club soda
• Rub club soda into the spot. Clean up with a sponge.

291. CORNSTARCH CLEANER
1 to 2 tablespoons cornstarch
• Cornstarch will absorb the stain. Pour some cornstarch onto the stain, let rest, then vacuum thoroughly.

292. PERFUME AND OTHER HOUSEHOLD CHEMICAL CLEANER
baking soda
• Cover the rug with baking soda until it is white. Let the baking soda stay on at least overnight. Vacuum thoroughly. If the odor lingers, have the rug steam cleaned as described in the introduction to this chapter.

293. VINEGAR STAIN BE-GONE
¼ cup vinegar, ¼ cup water
• Combine ingredients and rub with a sponge onto the stains. Rinse well with water.

294. SOAP AND VINEGAR
1 teaspoon vegetable-oil-based liquid soap, ¼ cup vinegar, ½ cup water
• Put ingredients in a jar and shake well. Saturate a sponge with the mixture and rub onto the stains. Rinse well.

295. ALCOHOL FOR STAIN REMOVAL
vodka or other plant based alcohol, vinegar
• Rub the stain with alcohol and rinse with vinegar.

Chapter 14

WALLS AND WALLPAPER

There are a medley of approaches available for cleaning walls. Chapter 2, All-Purpose Household Cleaners, Chapter 12, Floors, and Chapter 17, Stains and Tough Jobs, are a few of the chapters I suggest looking at that have formulas appropriate for cleaning walls. This chapter presents only a few of the many possibilities.

☐ **NATURAL INGREDIENT CHOICES**
Supermarket: bread, borax, washing soda
Health Food Store/Mail Order: vegetable-oil-based liquid soap
Hardware: chalk
Art Supply Store: soft brown eraser
☐ **CLEAN & GREEN COMMERCIAL PRODUCTS**
Health Food Store/Mail Order
A.F.M. Enterprises
   Super Clean
   Wallpaper Adhesive
See also Chapter 2, All-Purpose Household Cleaners, for brands of vegetable-oil-based liquid soaps.

---

296. WALLPAPER CLEANER I
1 or 2 slices of white bread
• Rub dirt and smudges on wallpaper away with the bread!

297. CHALK-STICKS WALLPAPER CLEANER
chalk sticks
• Rub the chalk lightly over stains and smudges. Brush clean.

298. SOAP-AND-WATER CLEANER
⅛ to ¼ cup vegetable-oil-based liquid soap, 2 gallons water
• Combine soap and water in a bucket. Stir until sudsy. Wash the walls with a sponge.

299. WALLS AND WALLPAPER
¼ cup borax, 1 gallon hot water
• Dissolve the borax in the hot water. Stir to blend. Saturate a sponge with the borax and water mixture and wash the walls.

300. HEAVY DUTY CLEANING FOR WALLS AND WALLPAPER
½ cup washing soda, 2 gallons hot water
• Dissolve the washing soda in the hot water. Stir to blend. Wearing gloves, saturate a sponge and wash the walls. Rinse thoroughly.

301. CLEANING WALLS WITH SHINY PAINT
½ teaspoon washing soda, 2 cups hot water, spray bottle
• Place the washing soda in the spray bottle and dissolve the washing soda with the water by gently shaking the bottle. Spray onto the walls and wipe dry with a clean cloth.

302. SMUDGE CLEANER
art store eraser (the brown crumbly kind)
• Rub the marks away.

303. VELVET WALLPAPER CLEANING FOAM
3 tablespoons vegetable-oil-based liquid soap, ¼ cup water
• Place the ingredients in a bowl and wisk into a foam. Scoop the foam onto a clean sponge and wash the wallpaper. Rinse well.

Chapter 15

FIREPLACE CLEANERS

□ **NATURAL INGREDIENT CHOICES**
Supermarket: vinegar, salt, washing soda, abrasive green pad
Mail Order: washing soda
Hardware: powdered graphite
Pharmacy: zinc oxide

---

304. THE-GREEN-PAD-AND-VINEGAR SOLUTION
¼ cup vinegar, green pad
• Scrub the vinegar onto the brick or stone with a mildly abrasive supermarket green pad. Rinse.

305. SOOT CLEANER
¼ cup washing soda, enough hot water to make a paste
• Dissolve the ingredients in a bowl, scoop onto a mild abrasive supermarket green pad, and scrub the affected area. Rinse. If there is a a heavy odor, leave the washing soda on until it dries before rinsing. This recipe requires a lot of rinsing. For less rinsing mix the washing soda with up to 2 gallons of water.

306. SMOKE RESIDUE REMOVER
⅛ cup salt, a few squirts of zinc oxide, enough water to make a paste
• Blend salt, zinc oxide, and water into a paste and rub onto the spot. Rinse well with water.

307. CAST IRON WOODSTOVE TOUCH-UP
powdered graphite
• Rub the graphite onto the area needing a touch-up.

Chapter 16

LAUNDRY

Laundry Soaps

Like most people, I enjoy the convenience of buying a laundry soap. I don't have the incentive to make my own soap flakes, even though I know they would work well and the process would not actually take long. Because I think I am not alone in my desire for an accessible, safe laundry soap, I have devoted most of this section to acceptable commercial products.

The commercial laundry soaps I recommend in this chapter:
1. Contain vegetable-oil-based soaps or detergents.
2. Biodegrade in a matter of days.
3. Do not contain EDTA (which binds with heavy metals in our waterways).
4. Do not contain cancer causing NTA.
5. Do not contain phosphates (which contribute to algae growth).
6. Do not contain bleach (which can become long lasting organo-chlorine compounds which in turn are stored in fatty tissue).
7. Do not contain non-biodegrading optical brighteners.
8. Do not contain synthetically derived preservatives.

Addresses of manufacturers and distributors of acceptable products are listed in Chapter 22.

□ NATURAL INGREDIENT CHOICES
Supermarket: washing soda, borax
Health Food Store/Mail Order: vegetable-oil-based liquid soap, perfume-free soap bar

□ CLEAN & GREEN COMMERCIAL PRODUCTS
Health Food Store/Mail Order

Cal Ben Soap Company
  Pure Soap
Dr. Bronner's
  Sal Suds
  Pure Castile Soap
Ecco Bella
  Suds Soap
  Laundry Booster and
  Whitener
Ecover
  Laundry Powder
  Laundry Powder
   Without Bleach
  Liquid Clothes Wash
  Wool Wash Liquid

Granny's Old Fashioned
Products
  Power Plus
Jurlique
  Gentle
Life Tree
  Premium Laundry
   Liquid
Naturally Yours
  Laundry Detergent
Simmons Pure Soaps
  vegetarian bar soaps
Tropical Soap Company
  Sirena Bar Soap

Supermarket

Arm & Hammer
  Super Washing Soda

Dial Corporation
  20 Mule Team Borax

LAUNDRY TIPS

Winter Washing Freeze Prevention: Add 2 tablespoons or more of salt to the rinse cycle to keep clothes from freezing stiff when hung outdoors to dry in winter.

Bleeding Dye Preventer: Add ¼ cup salt to the laundry load and rinse water. *Caution:* This recipe depends on the quality of the dye used. Test carefully.

Clothes Brighteners: Add 1 to 2 tablespoons epsom salts to the wash cycle or add 2 tablespoons powdered zeolite to the rinse cycle.

Note: I highly recommend not having your clothes dry cleaned, because the chemicals involved are indoor air pollutants, suspected carcinogens, highly toxic to both humans and the environment, and if that isn't bad enough, the chemicals used become hazardous waste.

308. IF YOU CAN FIND THEM! PERFUME-FREE LAUNDRY FLAKES

- It is not easy to find perfume-free soap flakes. If you do manage to locate them, buy extra because they have many uses for nontoxic cleaning, including use as a laundry soap.

309. DR. BRONNER'S
Sal Suds

A biodegradable soap with no phosphates, *Sal Suds* contains protein-bound castor oil, natural coconut and pine needle oils, potassium salts, and emollient skin conditioners.

Note: Pine is a common allergen.

310. ECCO BELLA
a.) *Suds Soap*

Contains castile soap, pine and coconut oils. This is an economical laundry soap because it is superconcentrated.

Note: Pine is a common allergen.

b.) *Laundry Booster and Whitener*

Made from sodium carbonate (washing soda).

311. ECOVER LAUNDRY PRODUCTS
a.) *Laundry Powder*

This coconut and palm oil-based soap biodegrades in five days. It contains no petroleum-based detergents or synthetic perfumes and is scented with natural citrus oils.

b.) *Laundry Powder Without Bleach*

Biodegrades in five days. No phosphates, no petroleum.

c.) *Liquid Clothes Wash*

This is a liquid laundry soap with similar ingredients to *Laundry Powder* but contains ethanol (derived from sugar, not petroleum). You only need 3 capfuls per load.

d.) *Wool Wash*

This coconut oil-based soap is appropriate for all natural fabrics. Fully biodegrades in three to five days.

312. GRANNY'S OLD FASHIONED PRODUCTS
Power Plus

This laundry concentrate is made of coconut oil-based soap.

313. JURLIQUE
Gentle

According to the company, this laundry soap contains: "Spagyrically processed herbs (soapwort, quillaya bark, horse chestnut) in soy whey; sodium lauryl sulphate, water, alcohol SD-40, coco betaine (derived from coconut oil), orange oil."

314. LIFE TREE
Premium Laundry Liquid
Ingredients are coconut-derived surfactants, natural oils, and a preservative derived from natural gum benzoin. You need as little as ⅛ cup of this highly concentrated liquid for a full load.

315. WASHING CHILDREN'S CLOTHES
recommended laundry soap, ¼ cup washing soda, ¼ cup borax
• A baby in diapers and learning to eat gets clothes thoroughly dirty. Adding a few soap boosters helps and this formula has been successful for many people. Follow manufacturer's directions for the laundry soap and simply add the washing soda and borax when you add the soap.

316. JAYA'S CASTILE AND ROSEMARY CLOTHES WASH
1 oz. liquid castile soap, 1 cup strong, cooled, rosemary tea
• Make the tea by steeping 1 teaspoon dried, crushed rosemary in 1 cup boiling water for 10 minutes. Let cool. Place the clothes in the washing machine and add the cooled, strained tea and castile soap. Wash as usual.
Note: Do not use on wool clothes.

317. LAUNDRY SOAP BOOSTER
½ cup washing soda
• Add ½ cup or more washing soda to your laundry, depending on how dirty the load is. Some commercial laundry detergents contain up to 68 percent washing soda!

318. LAUNDRY PEELS
bar of perfume-free soap, washing soda (optional), borax (optional)
• You can make your own "flakes" by grating soap with a grater and using these in your washing machine. Try to make the smallest peels possible so they will dissolve well. A little trial and error might be needed to get the right amount. Add washing soda (very effective against oil and grease) and borax (a water softener), and you have a very good washing concoction. You will need to wash with hot water.

319. WOOL, SILK, NATURAL FIBERS

2 teaspoons or less vegetable-oil-based liquid soap, water

• Put the soap into a sink or pan, add some cold water, and agitate to produce suds. Add clothing and cover with more cold water. Wash gently. Rinse well.

For wool: Make certain you wash wool in cold water or it will shrink. After washing, lay the items (one at a time) on top of a clean, dry towel and carefully roll up the towel tightly until it is shaped like french bread. "Walk" on the towel with your knees to squeeze out excess water. Unroll the towel and lay the garment out flat on a clean, dry towel. Reshape if necessary.

Stain-Removing Pastes and Presoaks

With all presoaks and pastes, test for colorfastness first by trying a small amount of the paste or liquid on an inconspicuous area of the clothing. See also Chapter 17, Stains and Tough Jobs.

□ NATURAL INGREDIENT CHOICES

Supermarket: washing soda, salt, baking soda, cream of tartar, borax
Health Food Store: shampoo, sodium percarbonate
Mail Order: sodium perborate (see Yellow Pages directory for a chemical supply company).
□ **IMPLEMENTS:** Use an old toothbrush or fingernail brush to scrub pastes into stains.

INGREDIENT HIGHLIGHTS

Mineral Magic: Custom-make your own stain removing pastes from the following minerals. The more minerals you use, the more they will work, but the more you have to rinse.

Baking soda: odor absorbing, deodorizing (don't use on wool or in aluminum)
Borax: disinfects, deodorizes, inhibits mold
Pumice: stain remover
Washing soda: removes grease

320. WASHING SODA PASTE

2 tablespoons washing soda, water

• Add small amounts of water to the washing soda until you have a paste. Rub the paste onto the spots and let it dry completely before putting the clothes into the washing machine.

Note: Arm & Hammer recommends wearing gloves when using washing soda pastes.

321. RING AROUND THE COLLAR
shampoo
• To remove "ring around the collar," rub a little shampoo, undiluted, into the stained cloth and then launder as usual.

322. SALT FOR PERSPIRATION STAINS
4 tablespoons salt, water
• Add enough water to salt to make a paste. Rub paste into the cloth. Let sit for an hour or so before laundering as usual.

323. BAKING SODA STANDBY FOR PERSPIRATION STAINS
4 tablespoons baking soda, water
• Add water drop by drop to baking soda until you have a thick paste. Rub into the cloth and let it sit for an hour or so. Wash as usual.

324. CREAM OF TARTAR FOR STAINS
2 tablespoons cream of tartar, water
• Make a paste of cream of tartar and water and rub onto the stain. Let the paste dry on the fabric before putting it into the washing machine. Launder as usual.

325. STAINS ON WHITE CLOTHES
2 tablespoons sodium percarbonate, water
• Mix sodium percarbonate with water until you have a paste. Work it into the stain and wash as usual.

326. SODIUM PERBORATE BLEACH
2 tablespoons sodium perborate
• Follow directions for Stains on White Clothes, above.

327. RIDDING CLOTHES OF THE "NEW" SMELL
baking soda
• This presoak is for low-level chemical contamination from normal household substances such as perfume, or for the removal of the smell of newly purchased clothing, not for industrial

chemical contamination. To rid clothes of low-level chemical smells, place the clothes in the washing machine with enough water to cover. Sprinkle one small-sized box of baking soda into the machine and agitate for a few minutes to dissolve and blend the baking soda. Soak the clothes overnight. When convenient during the soaking, agitate the machine for a few minutes. Launder as usual. Repeat this method until the clothes don't smell anymore.

328. DIAPER PRESOAK
½ cup borax, 1 diaper pail of hot water
• Cloth diapers are not nearly as hard to cope with as some would lead you to believe. If you use a cloth cover like a *Nikky*, they are as easy to put on as a plastic diaper and they are recyclable. Fill the diaper pail with hot water and add ½ cup of borax. Soak the diapers in this solution until you wash them. (Commercial diaper deodorizers can contain toxic ingredients.)

329. SALT-WATER SOAK FOR PERSPIRATION STAINS
¼ to ½ cup salt, enough water to cover clothes in washing machine
• Soak clothes in salt water for an hour or two. Wash as usual.

330. VINEGAR OVERNIGHT SOAK FOR PERSPIRATION STAINS
¼ cup vinegar, enough water to cover clothes in washing machine
• Add vinegar and water to the washing machine. Agitate to blend. Add clothes, agitate, and let sit overnight. Wash as usual.

Water Softeners

Hard water is high in mineral content. Although hard water is supposed to be good for the heart, it can be tiresome for people who live with it because it not only leaves difficult-to-clean mineral deposits on fixtures and appliances, but it also can turn clothes a grayish hue. Soap doesn't form suds as well in hard water, nor do cleaning-type minerals such as baking soda and washing soda perform as well. Adding a water softener to your laundry, however, helps solve a number of these problems at once.

☐ **NATURAL INGREDIENT CHOICES**
Supermarket: baking soda, borax, vinegar, washing soda
☐ **CLEAN & GREEN COMMERCIAL PRODUCTS**

<u>*Health Food Store/Mail Order*</u>

G&W Supplies
  Odor-fresh Zeolite

<u>*Supermarket*</u>

Arm and Hammer   Dial Corporation
  Baking Soda   20 Mule Team Borax
  Super Washing Soda

331. LOTS OF OPTIONS
• To increase the performance of soap, add any of the following ingredients to the wash cycle. To protect against clothes becoming grayish, add the same amount of ingredient to the rinse cycle as well. To a full load add:

½ cup baking soda
¼ cup vinegar
1 tablespoon Odor-Fresh Zeolite
¼ cup borax
½ cup washing soda

Laundry Bleach

As with presoaks, test these alternative bleach recipes on inconspicuous areas of clothing.

☐ **NATURAL INGREDIENT CHOICES**
Supermarket: washing soda, vinegar, baking soda, lemons
Health Food Store/Mail Order: sodium perborate (look in the Yellow Pages under chemical supply company), sodium percarbonate
☐ **CLEAN & GREEN COMMERCIAL PRODUCTS**

<u>*Health Food Store/Mail Order*</u>

Ecover   G&W Supply
  The Alternative Bleach   Odor-Fresh Zeolite
                                    Powder

332. BLEACH FOR WHITE CLOTHES
1 to 3 tablespoons sodium percarbonate per load

• Use 3 tablespoons sodium percarbonate if you need to remove stains such as tea, coffee, or berries. Add to wash cycle.

333. WASHING-SODA BLEACH
¼ cup washing soda
• Add washing soda to wash cycle.

334. FOR GREYING LAUNDRY
¼ cup vinegar
• Add vinegar to wash cycle.

335. SODIUM PERBORATE FOR STAINS
3 tablespoons sodium perborate
• Add sodium perborate to wash cycle.

336. FABRIC WHITENER
¼ cup borax
• Add borax to wash cycle.

337. LEMON JUICE BLEACH
¼ cup lemon juice
• Add lemon juice to wash cycle.

338. GREY AWAY
2 or 3 tablespoons zeolite powder
• Add zeolite powder to rinse cycle.

339. SUN
• Hang your clothes in the sun to dry. The sunshine is a natural bleach.

340. SUN AND LEMON JUICE
¼ cup lemon juice, sun
• Put lemon juice in the rinse cycle. After the clothes have spun dry, hang to dry in the sun.

Fabric Softeners

☐ NATURAL INGREDIENT CHOICES
Supermarket: baking soda, vinegar, borax
☐ CLEAN & GREEN COMMERCIAL PRODUCTS
Health Food Store/Mail Order

Ecover
    Fabric Conditioner

Naturally Yours
    Laundry Detergent
    Natural Bleach and
    Softener

---

341. BAKING SODA SOFT AS SNOW
¼ cup baking soda
• Add baking soda to wash cycle.

342. VINEGAR FOR FABRIC SOFTENER AND ANTI-CLING
¼ cup vinegar
• Add vinegar to wash cycle.

343. BORAX FABRIC SOFTENER
¼ cup or less borax
• Add borax to wash cycle.

Mold and Mildew Removal

344. Rx FOR MOLD: BORAX
⅛ to ¼ cup borax, water to cover
• Soak the clothes in a borax and water solution until the mold is gone. Wash as usual. For more on mold stains, see Chapter 17, Stains and Tough Jobs.

345. ZEOLITE FOR MOLD
1 tablespoon zeolite powder
• Add zeolite to wash cycle.

346. VINEGAR MOLD ATTACKER
¼ cup vinegar
• Add vinegar to the wash cycle.

Laundry Starch

347. STARCH SPRAY
2 to 3 teaspoons cornstarch, 1 cup water, spray bottle
• Combine cornstarch and water in a spray bottle. Shake well. Proceed as with commercial starches.

348. STARCH FOR DARK CLOTHES
½ cup cooled black tea, Starch Spray (above)
• Follow directions for Starch Spray, adding the cooled tea. The dark tea keeps the light-colored starch from showing on dark clothes.

Final Rinse

349. ELIMINATE CHEMICAL RESIDUE
¼ cup baking soda or vinegar
• Adding one of these ingredients into the final rinse of your laundry will help to eliminate a commercial detergent's chemical residue.

350. NO STATIC CLING
¼ cup vinegar
• Add vinegar to the rinse cycle.

351. NO LINT
¼ cup vinegar or lemon juice
• Add to the rinse cycle.

Chapter 17

STAINS AND TOUGH JOBS

□ NATURAL INGREDIENT CHOICES

Supermarket: meat tenderizer, salt, vinegar, milk, soda water, cornstarch, hydrogen peroxide, cornmeal, egg white, borax, baking soda, toothpaste, oil, bread, bran, unscented kitty litter, lemons, rubbing alcohol, Crisco, potatoes, rhubarb, tomatoes, cream of tartar, washing soda

Health Food Store: vegetable-oil-based liquid soap, sodium perborate, vegetable glycerin, Australian tea tree oil

Mail Order: washing soda, diatomaceous earth

Garden: potatoes, rhubarb, tomatoes

□ CLEAN & GREEN COMMERCIAL PRODUCTS

Health Food Store/Mail Order

A.F.M. Enterprises
  Super Clean
  Safety Clean
  X158
Auro Organics
  Cleansing Emulsion
  Plant Thinner

Granny's Old Fashioned
  Products
    Soil Away
  Greenspan
    Friendly Cleaner
  Naturally Yours
    Natural Solvent Spotter

Supermarket

Arm & Hammer
  Baking Soda
  Super Washing Soda

Dial Corporation
  20 Mule Team Borax

Stains

The following recommendations for stain removal are in alphabetical order by stain. Since each stain situation is unique, I suggest you make up your quantities according to the job. (One teaspoon of cornstarch will not help you if the stain is three feet wide.) As with all formulas, try a test strip first. It is always best to clean a stain as soon as possible, when it is still fresh and wet. *Note:* Make sure to rinse well with water after applying stain remover.

352. BABY FORMULA Rub with unseasoned meat tenderizer.

353. BERRIES AND RED WINE Soak in white wine.

354. BERRIES Pour boiling water from a height of three feet onto taut fabric.

355. BERRIES Soak in vinegar.

356. BERRIES Soak the stained cloth in milk.

357. BERRIES Boil nonshrinkable cloth in milk in a non-aluminum pot.

358. BERRIES Rinse with soda water.

359. BERRIES Rub vegetable glycerin into the stain before washing.

360. BERRIES Washing soda paste rubbed into stain.

361. BLOOD Make a paste of cornstarch and water and rub into the stain.

362. BLOOD Wipe with hydrogen peroxide.

363. BLOOD Rub with a cornmeal or cornstarch and water paste.

364. BLOOD Rinse with club soda.

365. BUTTER Make a paste of washing soda and water and rub into the stain.

366. CHEWING GUM Place clothing in freezer, or freeze the gum with ice cubes. Then pull gum off.

367. CHEWING GUM Rub full-strength vinegar onto the gum.

368. CHEWING GUM Cover with egg white; loosen the gum.

369. CHOCOLATE AND COCOA Mix sodium perborate and water to a paste, rub on the stain, then launder as usual.

370. CHOCOLATE Wipe with hydrogen peroxide.

371. CHOCOLATE Borax and water paste rubbed onto the stain.

372. CHOCOLATE Washing soda and water paste rubbed onto the stain.

373. CHOCOLATE Rub with any vegetable oil.

374. COFFEE AND TEA Pour boiling water from a height of three to four feet onto taut fabric.

375. COFFEE AND TEA Mix sodium percarbonate and water to a paste and rub onto the stain.

376. COFFEE AND TEA Soak in strong vinegar solution.

377. COFFEE AND TEA Simmer cloth (preshrunk only) in milk to cover in a non-aluminum pan.

378. COFFEE AND TEA Pour soda water on stain, then rinse.

379. CRAYONS Rub the marks with baking soda using a mild abrasive pad like a supermarket green pad.

380. CRAYONS Rub with toothpaste.

381. CRAYONS Rub with baking soda and olive oil paste. Recommended on washable surfaces only, because the oil can cause a stain of its own. I like to use this cleaner on walls.

382. CRAYONS Make a paste of washing soda and water. Rub onto the marks until gone.

383. DECALS Rub the decal with vegetable oil.

384. DECALS Rub with vinegar.

385. DECALS Rub with oil and vinegar salad dressing without herbs.

386. EGG Wash in cold water.

387. FOOD Rinse with club soda.

388. FRUIT AND JUICES See Berries.

389. GRASS Soak in vinegar.

390. GRASS Soak in a strong alcohol and water solution.

391. GREASE Rub with bread.

392. GREASE Cover with wheat bran, rub in and wipe off.

393. GREASE Absorb with cornmeal and wipe off.

394. GREASE Cover with borax, rub in and wipe off.

395. GREASE Cover with cornstarch, rub in and wipe off.

396. GREASE Cover with unscented kitty litter until the kitty litter has absorbed the grease; wipe off.

397. GREASE Cover spot with pure potato broth; rinse off.

398. GREASE Make a washing soda and water paste. Rub into the grease.

399. GREASE Rub with a dry bar of soap.

400. GREASE Cover the grease with salt, rub in and wipe off.

401. GUM LABELS Scrape off what you can with your fingernail. Dab label with vegetable oil. Let sit until glue softens and scrape remainder off.

402. **INK** Soak stain in lemon juice.

403. **INK** Make a paste with sodium perborate and water; rub into the ink stain.

404. **INK** Make a milk, vinegar and cornstarch solution and soak the stain.

405. **INK** Soak in milk.

406. **IODINE** Wipe with alcohol.

407. **LIPSTICK** Rub with white supermarket-type toothpaste.

408. **LIPSTICK** Rub with olive oil and baking soda paste. Recommended for washable surfaces only as the oil can leave a stain of its own.

409. **LIPSTICK** Rub with paste of washing soda and water.

410. **MACHINE OIL** Rub with vegetable or nut oil.

411. **MACHINE OIL** Saturate with kitty litter.

412. **MACHINE OIL** Scrub with washing soda and water paste.

413. **MEAT STAINS** Soak in strong washing soda and water solution.

414. **PETROLEUM OIL** Saturate with diatomaceous earth.

415. **MOLD** Rub with Australian tea tree oil.

416. **MOLD** Rub with borax and water paste.

417. **MOLD** Soak in strong vinegar solution.

418. **MOLD AND MILDEW** Rub with lemon juice and salt.

419. **MUSTARD** Rub with vegetable glycerin.

420. **NAIL POLISH** Rub with alcohol.

421. OIL Wipe with vegetable oil (oil draws out oil).

422. PAINT Soak in milk.

423. PAINT Soak in hot vinegar.

424. PAINT Soak in water and washing soda.

425. RUST Rub with cooked, cooled rhubarb. Wash off.

426. RUST Wash with lemon juice.

427. RUST Wash with alum and lemon juice paste.

428. RUST Make a paste of lemon juice and salt; rub on rust stains.

429. RUST Wash with alum and vinegar paste.

430. RUST Rub with tomatoes.

431. SHELLAC Wipe with alcohol.

432. STAINS Rub with a paste of borax and vinegar.

433. STAINS Soak in equal parts milk and vinegar. Wash off.

434. STAINS Rub with a mixture of 1 part vegetable glycerin and 2 parts vegetable-oil-based liquid soap.

435. STAINS Wash with a washing soda and water paste.

436. WAX Rub with paste of washing soda and water.

437. WINE See Berries.

Tough or Unusual Jobs

438. CHALKBOARD CLEANER
¼ cup vinegar, ⅛ to ¼ teaspoon vegetable-oil-based liquid soap, 1 cup water
• Mix ingredients together in a bowl, saturate a sponge or cloth with the mixture and wash the blackboard. Rinse well.

439. PIANO KEYS
3 tablespoons rubbing alcohol, ¼ cup water
• Combine ingredients and wipe the piano keys using a soft cloth. No need to rinse.

440. HAIR BRUSH CLEANER
¼ cup vinegar, water as needed
• Soak hairbrush in a bowl with the vinegar and enough water to cover. After soaking overnight, the grime can effortlessly be wiped out using a clean rag or swab.

441. SKUNK SMELL
tomatoes or tomato juice
• Our dog got sprayed by a skunk once and we didn't have any tomato juice. Our garden was full of ripe tomatoes, so I squashed tomatoes all over her, and yes, the skunk smell did go away. However, the dog shook herself before I had her rinsed and the walls and floor were covered with tomato. It looked like a blood bath! I suggest using only juice (and applying it outdoors).

442. SKUNK SMELL II
¼ cup dry mustard, water
• Mix mustard with water to a paste and cover the impregnated items. Leave on until dry. Rinse well.

Chapter 18

CARS: CLEANING, WAXING, RUST PROTECTION

□ **NATURAL INGREDIENT CHOICES**
Supermarket: unperfumed soap flakes, washing soda, aluminum foil, lemon juice, baking soda, cider vinegar
Health Food Store: food-grade linseed oil, vegetable-oil-based liquid soap
Mail Order: washing soda, carnauba wax, beeswax, zeolite
Art Store: beeswax
□ **CLEAN & GREEN COMMERCIAL PRODUCTS**
<u>Health Food Store/Mail Order</u>

A.F.M. Enterprises
  Vinyl Block
  Super Clean
Auro Organics
  Cleaning Emulsion
  Plant Thinner
Bon Ami
  Cleaning Powder
  Cleaning Cake

Greenspan
  Friendly Cleaner
G&W Supply
  Odor-Fresh Zeolite
Naturally Yours
  Degreaser
  Natural Solvent Spotter

Cleaning Cars

Note: See also Chapter 11, Furniture, for more recipes on cleaning upholstery.

443. SOAPY SUDS FOR THE CAR
¼ cup vegetable-oil-based liquid soap, bucket of warm water
  • Place the soap in a bucket and add water, swirling around to get it all sudsy. Saturate a sponge with the warm, soapy water and wash the car. If you have a hose, rinse with that. Otherwise, rinse with a bucket of fresh water and a clean sponge.

444. NATURAL AIR CLEANER

2 or 3 bags of *Odor-Fresh Zeolite*
- Place the zeolite in the car to absorb chemical odors.

445. CLEAN LIGHTS FOR NIGHT (HEADLIGHT CLEANER)

¼ cup baking soda, a squirt of liquid soap, enough water to make a paste
- Rub the paste onto the headlights with a sponge. Rinse thoroughly.

446. VINYL UPHOLSTERY CLEANER

1 teaspoon to ¼ cup washing soda, 1 cup boiling water
- Dissolve the washing soda completely with the boiling water. Saturate a sponge with the mixture and wash the vinyl. The more washing soda, the more thorough rinsing is required.

447. NEW-SMELL VINYL CLEANER

unperfumed soap flakes or vegetable-oil-based liquid soap
- Washing seems to clean the chemical smell from the surface of the vinyl, but it needs to be repeated frequently.

448. NEW-VINYL-SMELL ELIMINATOR

A.F.M. Vinyl Block
- *A.F.M. Vinyl Block* is a low-toxic sealant specifically designed to eliminate the smell of new vinyl. Follow manufacturer's instructions.

449. RADIATOR CLEANER

½ cup washing soda, 2 cups boiling water
- Dissolve washing soda in a bowl with boiling water. Using a sponge or brush, scrub the mixture onto the radiator. Rinse off with hot water and a rag.

450. NONTOXIC ENGINE DEGREASER

¼ cup *Plant Thinner*, ¼ cup baking soda, ¼ cup washing soda, 2 cups boiling water
- Put the dry ingredients in a bowl and cover with the boiling water, stirring to dissolve completely. You can try pouring the dissolved mixture into a spray bottle and spraying the minerals on the engine, but the sprayer may clog up. Instead of using a spray bottle, sprinkle the solution on the engine as well as you can. Wipe the loosened grease off with rags.

451. TAR REMOVER
food-grade raw linseed oil
• Wet a rag with linseed oil and rub the tar until it is gone.

452. TAR AND GREASE REMOVER
Plant Thinner or *Friendly Cleaner*
• Rub the solvent onto the tar until loosened.

Cleaning Chrome

453. CHROME CLEANER
¼ cup baking soda, enough water to make a paste
• Scoop the paste onto a sponge and rub the chrome. The baking soda gives good abrasive cleaning power for spots and stains. Once the chrome is cleaned, rinse well with warm water and polish. This formula is very effective.

454. LEMON JUICE ON CHROME
juice of 2 or 3 lemons
• Saturate a cloth or a sponge with lemon juice and wash the chrome.

455. LEMON RIND FOR CHROME
1 lemon, sliced in half and squeezed
• Holding onto the outside of the lemon, rub the chrome until it is clean.

456. VINEGAR
¼ cup cider vinegar
• Saturate a sponge with vinegar and rub the chrome until it is clean.

Waxing

If you are like me, you probably thought car wax had to be made in a factory. I have discovered this is not only false, but you can learn to custom-make waxes to suit your needs. For example, carnauba wax is one of the hardest waxes there is, so the more carnauba in your recipe, the harder the finished wax will be. But, if the wax is too hard, it will be difficult to rub onto the car. My formula for Annie's Own Car Wax makes a wax that, once it solidifies, is a bit harder than a paste wax. It rubs onto the metal beautifully and is totally nontoxic. Beeswax gives a very rich, characteristic fragrance to the wax, but you may add a fragrant herb or a few drops of an essential oil if you prefer.

INGREDIENT HIGHLIGHTS

Linseed Oil: Although linseed oil is used in many formulas because it has a fast drying time, linseed-oil products available from hardware stores contain petroleum products to speed drying time even more. I feel strongly enough about buying pure oil to suggest buying it in a health food store instead of a hardware store. When heated, an impure oil will emit toxic gases which could volatilize and ignite. In health food stores, raw linseed oil is sometimes known as the "Omega 3" oil. Petroleum-free boiled linseed oil (boiled linseed oil dries faster) is available from Auro Organics.

Turpentine Alternative: With hesitation I have included a few recipes in which the original formula required turpentine as a solvent. I have substituted *Plant Thinner* (made by Auro Organics) as an alternative product derived from citrus peels. The chemically sensitive should test this product carefully before using. *Note: Plant Thinner* is flammable.

457. ANNIE'S OWN NONTOXIC CAR WAX AND RUST PREVENTION

1 cup food-grade linseed oil, 4 tablespoons carnauba wax, 2 tablespoons beeswax, ½ cup vinegar, a few drops essential oil of your choice (optional)

• Put all the ingredients except essential oil in the top half of a double boiler (set over water) or saucepan. Heat slowly until the waxes are melted. Stir the ingredients to blend. Add the essential oil, blend, and pour into a heat-resistant container. After the wax has solidified, break it out of the container and rub on your car. Next, saturate a corner of a soft cotton rag with vinegar and buff and polish the wax to a high shine.

458. BEESWAX FOR YOUR CAR

4 oz. beeswax, 1 cup *Plant Thinner*

• Put the beeswax into the top half of a double boiler (over water) and heat slowly until it is melted. Stirring constantly and not leaving the pan unattended, add the *Plant Thinner* bit by bit until the beeswax and thinner are blended. Immediately take the mixture off the stove and let cool. Wax your car as you normally would.

Caution: Plant Thinner is flammable.

459. BEESWAX AND CARNAUBA WAX
4 tablespoons beeswax, 5 tablespoons carnauba wax, 1 cup linseed oil, 4 tablespoons *Plant Thinner*, a few drops of essential oil (optional)

• Put the beeswax, carnauba wax, and linseed oil in the top half of a double boiler (over water), cover, and melt the waxes slowly over a low heat. Blend well. Once the waxes are blended, add the *Plant Thinner* and essential oil. Stir constantly, not leaving the pan unattended, until thoroughly mixed. Pour into a heat resistant container. Cool. Rub the wax onto the car and polish with a clean, soft rag.

Caution: Plant Thinner is flammable.

Rust Protection

460. RUST PROTECTION
food-grade linseed oil

• Salting roads in the winter makes for serious rust problems, as any person who owns a car in the snow belt will tell you. Washing the salt off frequently is a help in rust prevention. Additionally, you can help protect against rust if you saturate a rag with linseed oil and polish the car with it.

461. RUST PREVENTION
1 cup food-grade linseed oil, ⅓ cup *Plant Thinner*

• Saturate a rag with the mixture and polish the car's metal.

Caution: Plant Thinner is flammable.

462. METAL ON METAL RUST REMOVAL
2 or 3 sheets aluminum foil, water

• Wet aluminum foil with water. Rub just the rust with the wet aluminum foil. Be careful, as the foil will scratch chrome.

Chapter 19

CLEANING UP CHEMICALS

Indoor air pollution is a significant problem in many of our homes (sometimes worse than outdoor pollution), and often it is caused simply by commonly used chemicals such as household cleaners, or fumes from perfume, wax or dry-cleaned clothes. Eliminating the problem can be as simple as cleaning with the right nontoxic solution. In this chapter there are some additional approaches that will help reduce existing pollution. The freer your house is of chemicals, the more often you will have a "breath of fresh air."

□ **NATURAL INGREDIENT CHOICES**
Supermarket: nondeodorized kitty litter, baking soda, washing soda, cornstarch, dry mustard, vinegar, lemon juice
Health Food Store/Mail Order: washing soda, food-grade linseed oil, vegetable-oil-based liquid soap
Art Store: potter's clay
□ **CLEAN & GREEN COMMERCIAL PRODUCTS**
<u>Health Food Store/Mail Order</u>

A.F.M. Enterprises
  Carpet Guard
  Sealants
  Super Clean
  Vinyl Block
  Water Seal

Auro Organics Natural
  Plant Chemistry
  Plant Thinner
Greenspan
  Friendly Cleaner
G&W Supplies
  Odor-Fresh Zeolite

INGREDIENT HIGHLIGHTS
Plants: Plants can be used as chemical air cleaners. NASA studies have shown that in closed situations plants can absorb certain chemicals (such as formaldehyde) from the air.
Washing Soda: I call this mineral, along with baking soda, the friend of the chemically sensitive. Washing soda is very effective at eating up petroleum-based products.
Zeolite: Zeolite is a one hundred percent nontoxic, naturally

occurring mineral found near volcanic activity. What makes zeolite unusual is that it is the only mineral negatively charged in its natural state, which means it naturally absorbs pollutants from the air. It is commonly used in water softeners.

Cleaning The Air

463. PLANT CHEMICAL AIR CLEANERS
• English ivy absorbs benzene, fig trees and spider plants absorb formaldehyde, and potted chrysanthemum and aloe vera absorb many toxins. The more plants, the better. Most effective in a closed area.

464. THE AMAZING CHEMICAL ODOR ABSORBER
baking soda
• Pour baking soda into open containers and place around the home. Most effective in small enclosed areas, like closets.

465. ZEOLITE CHEMICAL AIR CLEANER
breather bags of *Odor-Fresh Zeolite*
• Place zeolite breather bags in the area that needs air cleaning due to chemicals. Contact G&W Supply for their suggestions regarding particular problems.

466. SIMPLE CHEMICAL AIR CLEANER
vinegar
• Place vinegar in areas that smell of chemical paints and sealants.

467. COMMERCIAL AIR CLEANERS
• See Chapter 22, Safe Commercial Products, for a list of mail order companies. Many sources of air filtering equipment will also give free consultations as to which would best suit your needs. Chapter 23, Resources, will steer you towards professionals who can offer further advice.

Cleaning Up Chemicals

468. ENGINE OIL SPILLS
1 bag nondeodorized kitty litter, 3 tablespoons washing soda, ½ teaspoon or less vegetable-oil-based liquid soap, water
• Cover the engine oil with kitty litter, continually reapplying

until the oil has been completely absorbed. Put the saturated kitty litter aside for safe disposal. When you are down to the residue, dissolve the washing soda in two cups hot water. Add liquid soap. Scrub the oil residue with a scrub brush and soapy solution until clean. Rinse well.

469. ENGINE OIL REMOVER
diatomaceous earth
• Follow directions for Engine Oil Spills, above.

470. CORNSTARCH FOR ENGINE OIL
box of cornstarch
• Follow directions for Engine Oil Spills, above.

471. TAR CLEANER
equal parts linseed oil and lemon juice
• Wet a cloth with the mixture and rub on the tar until removed.

472. FLOOR-WAX REMOVER
washing soda, water
• Cover the floor with a thick washing soda paste. Let it dry. The wax will bubble up a bit and begin to peel off. Scrub well. Repeat as necessary. This will completely remove the wax if you use enough washing soda. Thorough rinsing is required.

473. THE AMAZING CHEMICAL ABSORBER, ONCE AGAIN
baking soda
• *For clothes:* Soak items in very strong baking soda and water solution; if the clothing cannot be soaked in water, sprinkle heavily with baking soda and leave on until the problem is gone. Most people do not use enough baking soda or leave it on long enough to get the job done.

474. COMMERCIAL WINDOW-WAX REMOVER
1 teaspoon vegetable-oil-based liquid soap, 1 cup water
• Mix the soap and water and wash the windows using a soft cloth. Rinse well.

475. TO REMOVE PETROLEUM-BASED PRODUCTS
¼ cup washing soda, ¼ cup baking soda, water
• For washable surfaces only. To remove perfume, dust

removers, and other chemicals that tend to linger on surfaces, cover completely with a mixture of baking soda, washing soda, and enough water to make a paste. Let dry completely, then rinse well. Repeat as often as necessary. Thorough rinsing is required for this recipe.

476. PERFUME REMOVER
3 teaspoons dry mustard, enough water to make a paste
• Rub the mustard paste onto washable surfaces that smell of perfume. Let the paste dry before rinsing completely.

477. GETTING RID OF THE NEW PLASTIC SHOWER-CURTAIN SMELL
sunlight
• Lay the shower curtain in the sun, turning every few hours. The chemicals will outgas and the worst of the smell will be gone in a day.

478. VERY OLD-FASHIONED PAINT-SMELL REMOVER
2 or 3 buckets of hay, 1¼ cups vinegar per bucket, water
• Fill a few buckets with fresh hay, vinegar, and water to cover. Place them around the painted room. Replace the hay every six hours until the smell is gone.

479. HOUSEHOLD CHEMICAL SPILLS OR STAINS
kitty litter, potter's clay, water
• This should not be used on industrial-strength chemical spills. Clean up as much of the chemical as you can with pure kitty litter or disposable diapers (these two materials don't cross react with chemicals). Cover the residue with a paste of potter's clay (available at art stores) and water. Leave the clay on the stain for 6 to 12 hours, remove, and then cover it again with fresh clay, continuing this process until the problem is gone.

480. TEMPORARY FORMALDEHYDE SEALANT
aluminum foil, aluminum tape
• If you have cabinets or furniture made of materials containing formaldehyde (plywood, pressed wood, etc.), you can eliminate fumes for a short time by wrapping them with aluminum foil and foil tape. Wrap completely, making sure all edges where aluminum foil meet are sealed. This method is a good temporary solution for a lot of chemical outgassing problems.

Chapter 20

REDUCING CHEMICAL RESIDUES FROM FRUITS AND VEGETABLES

A fact of twentieth-century life is that our food is coated with pesticide and herbicide residues from the chemicals sprayed in the fields, and with waxes and fungicides painted onto the produce before shipping.

I am sure that many of us would buy organic food to avoid these chemicals if it were available and affordable. Given the fact that we cannot always meet those two requirements, there is often no alternative but to eat commercial produce. You can help reduce your overall chemical intake a great deal, however, by washing off the chemical residue that is on your fruit and vegetables. Unfortunately, a lot of pesticides, herbicides, and fungicides used today are systemic. This means that the chemicals have been absorbed into the plant and have become part of it. No amount of washing will get rid of these chemicals. However, when you wash produce as directed here, you will at least be able to remove chemicals that adhere to the surface.

□ CLEAN & GREEN COMMERCIAL PRODUCTS
Health Food Store/Mail Order

Dr. Bronner's   Edcor
   Pure Castile Soap   Fruit Wash
   Sal Suds

See Chapter 2, All-Purpose Cleaners, for brand names of vegetable-oil-based liquid soaps and cleaning tips.

---

481. WASHING CHEMICAL RESIDUE FROM PRODUCE
1. Choose a glass bowl that you will use just for this job. I suggest glass because it cleans well. Plastic will get gummy. An alternative is to stop up your sink and fill it with water. The only

problem with the sink option is that you will have to scrub your sink after each cleaning.

2. Buy a good vegetable scrub brush. Natural bristle vegetable brushes are available at your health food store or by mail order.

3. Pesticides and waxes are oily and require soap for removal. Put a teaspoon or so of vegetable-oil-based liquid soap into the bowl with cool water. Stir until suds form.

4. Soak the fruit and vegetables in the soapy water for ten or fifteen minutes. Using the vegetable brush, scrub the produce (only where scrubbing is appropriate - skip your lettuce!) while holding the fruit under running water. Rinse everything well and set on a rack to dry. Wash the bowl or sink completely.

5. Make sure you have a good, workable vegetable peeler. Unless organic, produce that has been waxed should be peeled whenever possible.

482. CLEANING ORGANIC PRODUCE
salt, vinegar, water
• If you are fortunate enough to be able to buy organic produce then you may be faced with some bugs in it. Fill a bowl or sink with water and add a teaspoon or so of salt or vinegar. To flush out bugs, add the produce and let soak for a few minutes. The bugs will die and float off the produce.

Chapter 21

LEAVING NO TRACE: RECYCLING AND REDUCING OUR USE OF PLASTICS

I presume that we all have turning points in our lives when something happens to make us change a pattern of behavior. For me, I was jolted out of my inertia concerning plastic when I was on a whale watch, a commercial boat trip that takes people out to sea in the hopes of whale sightings. This boat had an environmentalist on board narrating events and educating us about different ocean ecosystems. He told us the story of Pegasus, a humpback whale that the research group he belonged to had been sighting and tracking for a number of years. Pegasus had a baby calf this particular year, and plastic had lodged in the baby's mouth, preventing it from nursing. The calf was starving to death. I was a nursing mother myself when I heard this story, and it completely broke my heart and changed forever my view of plastic. Fortunately, this story has a happy ending. The research group got close enough to the calf, finally, to cut away the plastic so that it could drink its mothers milk and survive.

Although the story of Pegasus had a happy ending, there are too many tragedies involving sea mammals who get caught in plastic and strangle to death. Plastic has been found in even remote areas of every ocean and sea on earth. Where does all of it come from? The plastic that Pegasus's baby was strangling on could have come from my house. It could have come from any one of our houses. Plastic does not biodegrade; our landfills are bursting with it and our oceans and forests are polluted with it.

In learning to eliminate plastic from your home, I suggest watching how plastic comes in and out of your life for a couple of weeks. In doing this myself, I found that most of the plastic we "used" was in the kitchen. I use quotation marks because I soon realized that about 75 percent of the plastic that came into our house was in packaging. I never used it, I just threw it away! The next 15 percent of plastic we used was in *our* packaging - sandwich bags, plastic wrap, plastic food containers, and of course, garbage bags. The last 10 percent was plastic parts of things like children's toys and spoon handles. I have learned to reduce our use of plastic drastically by simply choosing a non-plastic alternative. What plastic we do have, we recycle at our local recycling center.

483. ALTERNATIVES TO PLASTIC COMMERCIAL PACKAGING

1. Buying Unpackaged Food and Buying in Bulk

I recommend browsing in your local health food store to find products that can be substituted for plastic-packaged goods. You should be able to purchase many items you need in bulk and store them in glass or re-usable containers. Buying food in this manner can even save you money, because you aren't paying for the packaging. Another alternative is to start a neighborhood food co-op.

2. Using Plastic

Reuse as many plastic containers and bags as you can. Health food stores often provide plastic containers for such items as honey and peanut butter. They are sturdy containers and useful for all sorts of things. Furthermore, you can take the container with you the next time you shop and use it instead of a new one. You can even wash plastic bags and reuse those, too.

3. Eliminating Plastic Wrap

If you start recycling glass jars you will be amazed at how many glass containers come through your kitchen every week. You can store many items in them: leftovers, dried beans and grains, salad dressings - you name it, you can store it. Storing leftovers in glass jars practically eliminated the need for plastic wrap in our home.

4. Using Cellophane Bags or Waxed Paper

Instead of using plastic sandwich bags in lunch boxes, try cellophane bags or waxed paper. Both are cellulos-based products not plastic-based. Many health food stores carry cellophane as an alternative to plastic.

5. The Garbage Bag Problem.

I suggest calling your local recycling center to find out how recycling is done in your area. Once you have the routine down, recycling is easy, and you will be amazed at how little garbage you have left over that needs a garbage bag! Bob Walker of

EarthRight Institute in White River Junction, Vermont recommends a very obvious solution for that which can not be recycled (garbage). Put the garbage directly into a covered 30 gallon can without a bag. When full, take it to the dump, empty it, go home, wash the garbage can (with borax and water), and start over again. No plastic bag!

6. Bring Your Own Bag to the Supermarket

When you are buying produce at the grocery store, resist bagging it in plastic bags. If you don't bring your own bag with you, ask for paper bags instead of plastic for produce. Instead of using either plastic or paper at the check-out counter, bring your own canvas or string shopping bag.

484. AVOIDING PLASTIC PRODUCTS

1. Diapers

Using cloth instead of plastic diapers was so obvious I almost forgot to put it in here! To wash, simply fill a diaper pail with hot water and a half cup of borax and drop the dirty diapers into the pail to soak. (Try to find a metal diaper pail at a department store or yard sale.) Then, depending on how many diapers you have, you'll need to wash them only once or twice a week. I recommend buying cloth diaper covers because they are so easy to use (no pins!). At night, just double-diaper the baby and you are all set. Cotton is soft, clean, and gentle for your baby's young skin. A further bonus is that old cloth diapers make great dusters.

2. Shower Curtain

Buy a cotton shower curtain instead of plastic.

3. Toys

Children's toys are almost always plastic today. You can save money and keep plastic consumption down at the same time by buying plastic toys at yard sales. Yard sales are a great source of children's toys, so you should have no trouble keeping your children supplied with all the same items you could buy new. When you get home, just bring out a brush or toothbrush and some borax and wash them until they look like new.

4. Other Items

When you need to buy something, opt for the one with the least amount of plastic. Why buy a plastic spoon when there is a wooden one that is just as cheap and good?

485. RECYCLING

- Call your local landfill and find out where and how to

recycle plastic. Many plastic containers are recyclable. It is just as easy to drop a plastic item into a plastic recycling box, bag, or bin, as it is to drop it into the garbage pail.

Chapter 22

COMMERCIAL PRODUCTS THAT ARE SAFE FOR HEALTH AND SAFE FOR THE EARTH

The fate of our earth depends upon a rapid change in consumer habits. The manufacture of most commercial cleaning products is responsible for thousands of tons of hazardous waste being released into the environment each year. In order to have an effect on the manufacture of hazardous waste and to help reduce hazardous waste in the home, we must start using nontoxic and natural products. Every time I was tempted to compromise on the criteria I used in determining which products to include in this book, I had to remind myself that if I included a product containing, for example, synthetic preservatives or petroleum-based surfactants, I would slow down the process of change.

How do we determine what is an environmentally acceptable product and what isn't? What criteria should we use?

In order for a consumer product to be recommended in *Clean & Green*, it had to meet two broad, yet stringent, requirements. It had to be safe for health and safe for the earth. I have investigated the ingredients in all of the commercial products recommended and believe these products are among the purest consumers can buy. If I exclude a company it doesn't necessarily mean that their products are unacceptable, I simply may not know of them. However, I have waded through many Material Safety Data Sheets provided by companies, and as much as I would like to include more, acceptable products are still in the minority. Fortunately, I have found a number of items that are pure, naturally-derived, and effective. I cannot give enough praise to the companies that produce them, since there are so few who maintain such high standards.

In recommending natural materials I am not suggesting we regress to pre-industrial times. On the contrary, I am advocating sophisticated research into the capabilities of essential oils, plants, minerals, and other natural materials, particularly those that are renewable.

I admit it is not always easy to determine what is acceptable. Not only are manufacturers' claims confusing, but one feels one has to be a chemist to decipher the ingredients listed on the product - and that is when you are lucky enough to have ingredients listed! Many products also contain anonymous "inert ingredients." These can be highly problematic substances, yet we are not told what they are.

You can learn to make educated choices by understanding the processes by which we are polluting the earth. By doing this you start to see why products have either a beneficial or negative impact on our world and this, in turn, helps you to make your decisions.

Let's take a look at a typical furniture polish. It will be made from petroleum distillates and synthetic chemicals. How do you know this? A good clue is that it will be labeled with such words as DANGER, WARNING, CAUTION, or FATAL IF SWALLOWED. Knowing that the ingredients are toxic immediately alerts you to the fact that this product probably would not exist without smokestacks, toxic dumps, and other forms of production pollution. This, for starters, makes it unacceptable. Next, we bring the furniture polish home and pollute our living environments. The fumes from the polished furniture waft into the air and since most of us keep our cleaning products under the kitchen sink, every time we do the dishes we breath small amounts of the polish. Next, consider its packaging. Is the material recyclable? If so, and it has to be rinsed out prior to recycling, then you are adding toxins to the waste water stream. Alternatively, if we throw the furniture polish bottle into the landfill, any residue leaches into our ground water. The next time it appears, it is possibly in our tap water. The clue to the whole scenario was on the label of the bottle where it said DANGER, CAUTION, or WARNING.

To help you determine what makes a product acceptable, keep the following suggestions in mind when you shop for cleaning products.

AVOID TOXINS: They create poisonous waters and toxic dumps, and are responsible for many endangered and extinct species of wildlife. They can also be neurotoxic, carcinogenic, and central nervous system depressants. If they find their way into our drinking water they can cause chronic illness.

CHOOSE NATURAL AND RENEWABLE RESOURCES THAT ALSO BIODEGRADE: If we choose products that are natural to our planet such as plants and minerals, they will decompose in an ecological way when finally discarded. A laundry soap made with coconut oil, for example, biodegrades in a few days. The coconut is also sustainable and renewable. If you have to choose a product with synthetic ingredients, choose nontoxic, food-grade chemicals.

CONSIDER THE LANDFILL: Every time we buy a product we should ask ourselves how much of the packaging will go to the landfill. If it is a lot, try to find an alternative.

CONSIDER THE POPULATION: Buying a toxic, polluting, nonrecyclable product may seem inconsequential until you start to realize that hundreds of millions of people might be using it also.

Ingredients and Implements

ALCOHOL, ALTERNATIVE TO PETROLEUM BASED: Plant Alcohol Thinner, available from Auro Organics or Livos, is made from beets and potatoes. Another alternative could be vodka or whiskey.

Distributor: Auro Organics, Livos

ALUM: A mineral. Alum is generally used for pickling and as a natural dye fixative but it also has remarkable astringent properties.

Available in the herb section of supermarkets.

AUSTRALIAN TEA TREE OIL: The essential oil of the Australian tea tree. The oil is a broad spectrum fungicide and bactericide.

Available in health food stores under the brand Desert Essence.

BAKING SODA: Bicarbonate of soda. Baking soda is an effective mineral cleaning agent that also deodorizes.

Distributor: Arm & Hammer Brand

BEESWAX: Beeswax is made from a wax that honeybees secrete to make their combs. Art stores, and furniture- and instrument-making supply houses carry pure beeswax. Beeswax has a natural aroma all its own.

Mail Order Supplier: Wood Finishing Supply Co. Inc.

BORAX: A mineral of natural origin consisting of water, oxygen, sodium, and boron. In large doses, borax is toxic if consumed orally; the lethal dose for a 150-pound person is between one ounce and one pound. Borax has antiseptic, antifungal, and antibacterial properties.

Distributor: The Dial Corporation as 20 Mule Team Borax

CARNAUBA WAX: Carnauba wax is a natural wax made from the leaves of a Brazilian palm tree. It is the hardest natural wax known.

Mail Order Supplier: Wood Finishing Supply Co., Inc.

CASTILE SOAP: Castile soap is a mild soap originally made from olive oil but today castile soap can be made from other vegetable oils as well. For more soaps to choose from, call some of the allergy-related mail-order companies listed at the beginning of the section Mail Order Suppliers.

Manufacturers: Dr. Bronner's, Sirena Tropical Soap Company, Chef's Soap, Ecco Bella, Simmon's Pure Soaps

CELLOPHANE BAGS: Cellophane is made from regenerated cellulose, which is derived from plants. Cellophane bags are a good substitute for plastic bags.

Available in some health food stores and in a large number of mail order catalogues.

CITRUS-SEED EXTRACTS: Citrus-seed extracts act as natural preservatives in cosmetics and are increasingly used as antifungal, antibacterial, and disinfectant agents in other products. They are used internally as antiparasitic remedies. The purest form of citrus-seed extract available over the counter is *Paramycocidin*.

Manufacturer: NutriCology

CREAM OF TARTAR: The purified and crystallized bitartrate of potassium.

Available in the herb section of supermarkets.

DIATOMACEOUS EARTH: Diatomaceous earth is made from the skeletons of a class of algae called diatoms. When the diatoms die, their skeletons become converted to or impregnated with silica. Diatomaceous earth is used for cleaning and for nontoxic pest control. When diatomaceous earth is called for in a recipe in this book or for pest control, it is not the kind of diatomaceous earth commonly found in pool-supply stores and used in filters.

Mail Order Suppliers: Ecco Bella, EcoSafe Formulas

ESSENTIAL OILS: Essential oils are the essence of a plant's fragrance and are the basis of perfumes. The oil is extracted from the plant in various manners, the purest being steam distillation. Another method of extraction is to soak the plant matter in oil or alcohol.

Available in health food stores and from mail order suppliers.

GLYCERIN: Vegetable-based glycerin is made from glycerol, which is a part of all natural fats and oils.

Available in health food stores.

LANOLIN: The fats and oils found in sheep wool.

Available at pharmacies.

LAUNDRY SOAPS: Vegetable-oil-based laundry soaps are recommended in this book. See also, Soaps, All-Purpose Liquid, below.

Manufacturers or Distributors: Ecover, Granny's Old Fashioned Products, Dr. Bronner's, Ecco Bella, Jurlique, Life Tree

LINSEED OIL: Linseed oil is made from flax seeds. It is used for furniture and wood care because it is a natural oil that dries. Raw food-grade linseed oil from a health food store is recommended because most hardware-store-grade linseed oil contains petroleum products and dryers. Raw linseed oil will take a little longer to dry. Linseed oil is sometimes called the "Omega 3" oil. The only acceptable brand of petroleum-free boiled linseed oil (which dries even faster) currently available is manufactured by Auro Organics Plant Chemistry.

Distributor: Auro Organics

NATURAL VEGETABLE SCRUB BRUSHES: Available in health food stores and environmentally conscious mail order catalogues. These brushes are very useful for scrubbing chemical residues from produce.

Available in health food stores and from mail order suppliers.

PENNYROYAL: An essential oil that has flea-deterrent qualities. Pregnant women should not use this oil.

Available in health food stores and from mail order suppliers.

PLANT THINNER: A citrus turpentine used as a solvent.

Distributer: Auro Organics

PUMICE STONE: Cooled lava becomes porous because of the escaping gases. Ground up, this rock is called pumice stone and is commonly found in two or three gradations of powder or as a stone. It is mostly obsidian. Pumice is used as an abrasive, polisher, and stain remover.

Mail Order Supplier: Wood Finishing Supply Co., Inc.

ROTTENSTONE: Rottenstone is made from crushed up stone and is used as an abrasive.

Mail Order Supplier: Wood Finishing Supply Co. Inc.

SHELLAC: Pure shellac is derived from lac, a secretion of beetles, and is used to seal wood.

Mail Order Supplier: Wood Finishing Supply Co., Inc.

SOAPS, ALL-PURPOSE LIQUID: Vegetable-oil- instead of petroleum-based liquid soaps are recommended in this book. These soaps can be used for dishes, fine washables, and household cleaning and are a component of many nontoxic recipes. Liquid all-purpose soaps are often sold in health food stores. See also Castile Soaps.

Health Note: Although the chemicals DEA, MEA and TEA (among others) are suspected of causing the formation of carcinogenic nitrosamines in cosmetics, it is unclear if they do this in soaps. To be on the safe side, it is suggested that you add a couple of drops of liquid vitamin E (available at your health food store) per ½ gallon to all commercial vegetable-oil-based liquid soaps recommended in this book to protect against possible nitrosamine contamination. To make this task easy, just add the vitamin E to the bottle when you open it the first time. You do not have to do this at all if either vitamin E or vitamin C (ascorbic acid) are listed in the ingredients.

Distributors and Manufacturers: Auro Organics, Dr. Bronner's, Ecover, Granny's Old Fashioned Products, Infinity Herbal Products, Life Tree, Jurlique, Bau, Livos

SOAP FLAKES: Unscented soap flakes are hard to find. You can make your own by grating any recommended bar of soap.

SODIUM PERCARBONATE: A natural bleach alternative to chlorine made from washing soda and hydrogen peroxide.

Distributer: Ecover

SODIUM PERBORATE: A natural bleach alternative to chlorine made from borax and hydrogen peroxide.

Mail Order Supplier: Look in your Yellow Pages under chemical supply companies.

TURPENTINE ALTERNATIVE: See *Auro Organics Plant Thinner* or *The Greenspan Healthy Kleaner* later in this chapter under manufacturers.

VEGETABLE GLYCERIN: see Glycerin

VEGETABLE-OIL-BASED SOAPS: see Soaps, All-Purpose Liquid

VINEGAR: A liquid derived from the fermentation of fruits or grains. Its acid content makes it useful for cutting grease and dissolving mineral deposits.

Available in health food stores and supermarkets.

WASHING SODA: A mineral, sodium carbonate, also known as

soda ash and sal soda. A very effective cleaner of grease, oil, dirt, and many petroleum products.

Distributor: Arm & Hammer Brand

ZEOLITE: Zeolite is a one hundred percent nontoxic, naturally occurring mineral found near volcanic activity. What makes zeolite unusual is that it is the only negatively charged mineral in its native state, which means it naturally absorbs pollutants from the air. Zeolite is commonly used in water softeners.

Distributor: G&W Supply

Manufacturers and Distributors

AIR THERAPY: Essential oils distilled from real citrus, and nothing else. To be sprayed as a "purifying mist."

Mia Rose Products Dept. CG
177-F Riverside Ave.
Newport Beach, Ca 92663
(800) 282-6339

A.F.M. ENTERPRISES: Makes low-toxic products including many different kinds of cleaners, sealants, paints, waxes, mold- and mildew-control products, joint and spackling compounds, adhesives, carpet sealants, carpet adhesives, carpet soaps, and even shoe polish. The chemically sensitive tend to tolerate these products very well. A.F.M. also makes continuous-membrane products that successfully seal in toxic chemicals, and a soap called Super Clean that helps clean up chemicals.

Main Office:
350 West Ash Street, Suite 700, Dept. CG
San Diego, CA 92101
(619) 781-6860
Orders:
1960 Chicago Avenue, Suite E-7, Dept. CG
Riverside, CA 92507
(909) 781-6860

ARM & HAMMER: Arm & Hammer Baking Soda is available in most supermarkets in the bakery supply section, and their Super Washing Soda can be found in the laundry section.

Church & Dwight Co. Inc. Dept CG
P.O. Box 7648
Princeton, NJ 08543-7648

AURO ORGANICS NATURAL PLANT CHEMISTRY: This German company makes nontoxic, plant-derived, petroleum-free wood protectors, stains, paints, waxes, linseed oil, organic soaps, beeswax floor care, and furniture products.

Sinan Company, Dept. CG
P.O. Box 857
Davis, CA 95617-0857
(916) 753-3104

BIO/CHEM RESEARCH: Nutribiotic/Citricidal Brand of grapefruit seed extract. Products are available in health food stores.

Bio/Chem Research
P.O. Box 238, Dept. CG
Lakeport, CA 95433
(707) 263-1475

BIOFA: This German company manufactures household products such as petroleum-free and plant-derived Household Cleaner and Dishwashing Liquid. They also manufacture low-toxic building supplies. (The building supplies do contain petroleum.)

Bau Inc., Dept CG
P.O. Box 190
Alton, NH 03809
(603) 364-2400

BON AMI COMPANY: Manufactures Bon Ami Cleaning Powder and Bon Ami Cleaning Cake. Each of these specific products contain only feldspar and soap.

Faultless Starch/Bon Ami Co., Dept. CG
1025 West 8th St.
Kansas City, MO 64101

CAL BEN SOAP COMPANY: Makes rich and lathery bar soaps perfumed with natural almond oil.

C.B. Co. Manufacturers, Dept. CG
9828 Pearman St.
Oakland, CA 94603
(415) 638-7091

COASTLINE PRODUCTS: Distributor of soaps and detergents that are less toxic and environmentally preferable. Simple Soap and Coastline Old Fashioned Soap Products are a few of the products available.

Coastline Products, Dept. CG
P.O. Box 6397
Santa Ana, CA 92706
(800) 544-4112

DESERT ESSENCE: Distributor of pure Australian Tea Tree Oil, a broad spectrum fungicide.

Desert Essence, Dept. CG
9510 Vassar
Chatsworth, CA 91311

DIAL CORPORATION: Manufacturer of 20 Mule Team Borax.

Dial Corporation, Dept. CG
20 Mule Team Borax
Phoenix, AZ 85077-1717

DR. BRONNER'S: Manufacturer of pure castile liquid and bar soaps scented with essential oils. They also make a super-concentrated cleaner called Sal Suds. These natural soaps are available in most health food stores.

All-One-God-Faith, Inc., Dept. CG
P.O. Box 28
Escondido, CA 92033

EARTHRITE: Manufacturer of one of the best available laundry detergents without dyes, optical brighteners, perfumes, and other irritants.

EarthRite, Dept. CG
Corp. Center #1
55 Federal Road
Danbury, CT 06813
(203) 731-5000

EARTHWISE PRODUCTS COMPANY: Manufacturer of a wide range of off-the-shelf, nontoxic and environmentally preferable cleaning products. Available in health food stores.

Earth Wise Products Company, Dept. CG
Jersey City, NJ
07302-9988
(213) 011-8108

ECOVER PRODUCTS: A German company that produces bio-degradable household cleaning aids. Ecover's products are available from health food stores, food co-ops, supermarkets, and environmentally conscious mail order catalogues. Ingredients are all derived from natural materials. The chemically sensitive and allergic should test these products before using because they contain essential oils.

Mercantile Food Company, Dept CG
Carpenter Road
P.O. Box 55
Philmont, NY 12565
(518) 672-0190

G&W SUPPLY: G&W Supply is replaced by the Dasun Company as a source of a very pure form of zeolite.

Dasun Company, Dept. CG
P.O. Box 668
Escondido, CA 92033
(800) 433-8929

GRANNY'S OLD FASHIONED PRODUCTS: This company's products are EDTA- free. The detergents are made with coconut oil surfactants. Products include Granny's Aloe Care (dishwashing/all-purpose liquid soap), Granny's Power Plus

Laundry Concentrate, Soil Away (stain remover), and Karpet Kleen (carpet shampoo). Well tolerated by the chemically sensitive.

> Granny's Old Fashioned Products, Dept. CG
> P.O. Box 660037
> Arcadia, CA 91006
> (818) 577-1825

INFINITY HERBAL PRODUCTS: Maker of Heavenly Horsetail All-Purpose Cleaner. This detergent is available at most health food stores. The ingredients are all derived from natural materials.

> Infinity Herbal Products Dept. CG
> Division of Jedmon Products Ltd.
> Toronto, Canada M3J 3J9

KISS MY FACE: Manufacturer of a pure, olive oil, castile bar soap.

> Kiss My Face, Dept. CG
> P.O. Box 224
> Gardiner, NY 12525
> (619) 744-9680

LIFE TREE: These products are made from petroleum-free surfactants. All of the ingredients are derived from natural materials. Available in health food stores.

> Life Tree Products, Dept. CG
> A Division of Sierra Dawn
> P.O. Box 1203
> Sebastopol, CA 95472

LIVOS: This German company manufactures petroleum-free and plant-based cleaners. They also manufacture low-toxic building supplies. (Many of their building supplies do contain petroleum.)

> Eco-Divine, Dept. CG
> 1365 Rufina Circle
> Santa Fe, NM 87501

NATURALLY YOURS/ECOLO CLEAN: Manufactures a large selection of nontoxic cleaning products.

Ecolo-International, Ltd., Dept. CG
717 N. West Bypass
Springfield, MO 65802
417/865-6260

NUTRICOLOGY/ALLERGY RESEARCH GROUP: Manufacturer of Paramycocidin, which is a grapefruit-seed extract used internally to kill parasites and yeast. Recommended in its liquid form as a mold killer.

Allergy Research Group, Dept. CG
400 Preda St.
San Leandro, CA 94577

SEVENTH GENERATION: This company has worked hard and shown great commitment to provide us all with less toxic, more environmentally preferable products that work. They develop their formulas to high standards, and can now provide us with a superlative line for household cleaning. Available in health food stores.

Seventh Generation, Dept. CG
Colchester, VT 05446-1672
(800) 456-1198

SIMMONS PURE SOAPS: Manufacturer of hand-made castile and also 100% vegetarian bar soaps. Pure ingredients.

Simmons Handcrafts, Dept. CG
42295 Highway 36
Bridgeville, CA 95526

SUNFEATHER HANDCRAFTED HERBAL SOAP COMPANY: This catalog will introduce you to the world of soap — everything from the gourmet's delight to plain and simple soap blocks. The soap is beautifully packaged, often in fabric, and the catalog even offers soap making supplies. Highly recommended.

Sunfeather, Dept. CG
HCR 84
Box 60A
Potsdam, NY 13676

THURSDAY'S PLANTATION: Australian Tea Tree Oil products, broad spectrum fungicides.

> Thursday's Plantation, Dept. CG
> P.O. Box 5613
> Montecito, CA 93150
> (805) 963-3578

TROPICAL SOAP COMPANY: Their Sirena pure coconut-oil soap contains no perfume, deodorants, coloring, dye, animal fat or tallow, or synthetic detergents. Available in most health food stores. One of the most economical pure soaps.

> Tropical Soap Company, Dept. CG
> P.O. Box 797217
> Dallas, TX 75379
> (800) 527-2368

WYSONG CORPORATION: Manufacturer of Citressence, a "natural odor counteractant." Made from natural botanical extracts and essential oils.

> Wysong Corporation, Dept. CG
> 1880 N. Eastman
> Midland, MI 48640
> (800) 748-0188

Mail Order Suppliers

Many of the companies listed here serve the chemically sensitive and are therefore very familiar with the latest low toxic and nontoxic products. Many of the owners are chemically sensitive themselves. These companies are remarkably good resources and will give free consultations on nontoxic products, building supplies, water and air filters, and issues concerning indoor air pollution.

Allergy Relief Shop, Inc., Dept. CG
3371 Whittle Spring Rd.
Knoxville, TN 37917
(615) 522-2795

A.F.M. products, books, Extractionaire vacuum cleaners, etc. Free consultation.

Allergy Resource, Dept. CG (800)USE-FLAX	Bon Ami, Granny's Old Fashioned products, A.F.M. products, cellophane bags, etc.
The Allergy Store P.O. Box 2555, Dept. CG Sebastopol, CA 95473 Outside CA: (800) 824-7163 CA and Canada: (800) 950-6202	Granny's Old Fashioned products, A.F.M. products, etc. Free consultation.
An Ounce of Prevention 8200 E. Phillips Place Dept. CG Englewood, CO 80112 (303) 770-8808	Paramycocidin, allergy supplies, Granny's Old Fashioned products, pure organic linseed oil, etc. Special orders filled.
Baubiologie "Healthful" Hardware Catalog P.O. Box 3217, Dept. CG Prescott, AZ 86302-3217 (602) 445-8225	Advanced electromagnetic field information and protective supplies, A.F.M. products, books, chemical test kits, natural linoleum, radiation-free smoke detectors, full-spectrum lights, Granny's Old Fashioned products, etc.
The Dasun Company P.O. Box 668, Dept. CG Escondido, CA 92033 (800)433-8929	Distributors of a diverse line of zeolite products including "All-Natural Zeolite."
EcoSafe Products, Inc. P.O. Box 1177, Dept. CG St. Augustine, FL 32085 (800) 274-7387	Diatomaceous earth, nontoxic pet care.

EcoShop, Inc., Dept CG
5884 E. 82nd St.
Indianapolis, IN 46250
(317)84-WORLD

Nontoxic household cleaners, personal care products, paints, recycled supplies for home and office.

Environmentally Safe Products
8345 Walnut Hill Lane,
Ste. 225, Dept. CG
Dallas, TX 75231
(800) 428-2343

Cleaning materials, test kits and equipment, personal care products.

Heart of Vermont, Dept. CG
Route 132, Box 183
Sharon, VT 05065
802/763-2720

A company known for their organic wool bedding, they also sell some natural products including hair brushes, personal care products, soaps, and books.

Karen's Nontoxic Products
1839 Dr. Jack Road, Dept. CG
Conowingo, MD 21918
(800) 527-3674

Pure, natural products, including natural cleaning utensils, nontoxic cosmetics, cellophane bags, soaps, etc.

The Living Source
7005 Woodward Drive
Ste. 214, Dept. CG
Waco, TX 76712
(817) 776-4878

Major distributor of A.F.M. products, bar soaps, cellophane bags, nontoxic cosmetics, air cleaning units, etc. Free consultation.

The Natural Choice Catalog
1365 Rutina Circle, Dept. CG
Santa Fe, NM 87501
(505) 438-3448

N.E.E.D.S., Dept. CG
527 Charles Avenue
Syracuse, NY 13209
(800) 634-1380

Granny's Old Fashioned products, A.F. M. products, books, cellophane bags, nontoxic cosmetics, water and air filters, full spectrum lights, etc.

Nigra Enterprises
5699 Kanan Road, Dept. CG
Agoura, CA 91301
(818) 889-6877

A.F.M. products, Pace-Chem products, full spectrum lighting, water and air treatment products.

Nontoxic Environments
9392 S. Gribble Road
Dept. CG
Canbuy, OR 97013
(503) 266-5244

Air and water filters, nontoxic building supplies, books.

Real Goods
966 Mazzoni St., Dept. CG
Ukiah, CA 95482-3471
(800) 762-7325

Cleaning products, energy efficiency specialists, recycled papers, books.

Seventh Generation
49 The Meadow Park
Dept. CG
Colchester, Vt 05446
(800) 456-1177

Many cleaning products, recycled paper products, recycling equipment, energy saving light bulbs, etc.

Chapter 23

RESOURCES

ORGANIZATIONS GEARED TOWARD THE DEVELOPMENT OF HEALTHY, PRACTICAL AND ENVIRONMENTALLY PREFERABLE PRACTICES IN THE HOME

Citizen's Clearinghouse for Hazardous Waste, Inc.
P.O. Box 6806, Dept. CG
Falls Church, VA 22040

Environmental Health Coalition
1717 Kettner Blvd., Suite 100, Dept. CG
San Diego, CA 92101-2532

Household Hazardous Waste Project
1031 E. Battlefield, Ste. 214, Dept. CG
Springfield, MO 65807

Green Seal
1250 23rd St. NW, Ste. 275, Dept. CG
Washington, DC 20037-1101

Human Ecology Action League (H.E.A.L.)
P.O. Box 49126, Dept. CG
Atlanta, GA 30359-1126

Informed Consent/INRCH
P.O. Box 935, Dept. CG
Williston, ND 58802-0935

Mothers & Others
40 West 20th St., Dept. CG
New York, NY 10011

The Washington Toxics Coalition
4516 University Way NE, Dept. CG
Seattle, WA 98105

CONSULTATIONS ON NONTOXIC AND ENVIRONMEN-
TALLY SAFE CLEANING FOR INDUSTRY, BUSINESS AND
THE HOME

Carl Grimes
Healthy Habitats
1811 South Quebec Way #99
Denver, CO 80231
(303) 671-9653

Bergfeld's Housekeeping and Cleaning Services
P.O. Box 45
211 West 92nd St.
New York, NY 10025
(212)666-6649; (718) 670-7009

Jeff Innis
Healthful Cleaning Co.
1600 N. Willis #195
Bloomington, IN 47401
(812) 339-8770

CONSULTATIONS FOR NONTOXIC AND ENVIRONMEN-
TALLY SAFE BUILDING

Wayne Baltz
829 Gallup Road, Dept. CG
Ft. Collins, CO 80521

John Banta
Baubiologie Hardware
P.O. Box 3217, Dept. CG
Prescott, AZ 86302-3217
(602) 445-8225

John Bower
The Healthy House Institute
7471 North Shiloh Road, Dept. CG
Unionville, IN 47468
(812) 332-5073

Environmental Construction Outfitters
44 Crosby St., Dept. CG
New York, NY 10012

Mary Oetzel
Environmental Health Services
P.O. Box 92004, Dept. CG
Austin, TX 78757
(512) 288-2369

David Goldbeck
Smart Kitchen Associates
PO Box 87, Dept. CG
Woodstock, NY 12498
(914) 679-5573

Note: Most of the mail-order suppliers listed in Chapter 22 will also have information on the latest nontoxic building supplies. Many of them give free consultations. Don't be put off by the word "allergy" in many of the company names; their clients are the chemically sensitive, and there is no better source for information about chemical-free living than from people who have to live without chemicals.

SAFE FOR HEALTH, SAFE FOR THE EARTH

Pesticide Alternative Resources:
B.I.R.C. (Bio-Integral Resource Center)
P.O. Box 7414
Berkeley, CA 94707
(510) 524-2567

Environmental Protection Agency Pesticide Hotline
(800) 858-7378

National Coalition Against the Misuse of Pesticides
530 7th St., SE
Washington, DC 20003
(202) 543-5450

New York Coalition for Alternatives to Pesticides
P.O. Box 6005
Albany, NY 12206-0005
(518) 426-8246

Northwest Coalition for Alternatives to Pesticides
P.O. Box 1393
Eugene, OR 97440
(503) 344-5044

Organic Gardening
Rodale Press, Inc.
33 East Minor Street
Emmaus, PA 18049

For more on alternatives to miscellaneous chemicals we bring
into the house:

Alternatives
The Washington Toxics Coalition
4516 University Way NE
Seattle, WA 98105

Arts, Crafts and Theater Safety (ACTS)
181 Thompson St. #23
New York, NY 10012-2586

Citizens Clearinghouse for Hazardous Waste
P.O. Box 926
Arlington, VA 22216

Eco-Home Network
4344 Russell Avenue
Los Angeles, CA 90027

Environmental Building News
RR 1, Box 161, Dept. CG
Brattleboro, VT 05301
(802) 257-7300

The Green Guide for Everyday Life
Mothers & Others
40 West 20th St., Dept. CG
New York, NY 10011

The Human Ecologist
Human Ecology Action League (H.E.A.L.)
P.O. Box 49126, Dept. CG
Atlanta, GA 30359-1126

Informed Consent Magazine
P.O. Box 935, Dept. CG
Williston, ND 58802-0935

Natural Life Newsletter
The Alternate Press
272 Hwy #5, RR 1, Dept. CG
St. George, Ontario Canada NOE 1NO

Rachel's Environment & Health Weekly
P.O. Box 5036, Dept. CG
Annapolis, MD 21403-7036

Toxinformer
Environmental Health Coalition
1717 Kettner Blvd., Suite 100, Dept. CG
San Diego, CA 92101-2532

Information on Chemicals in our Foods:

Allergy Hotline Newsletter
P.O. Box 161132, Dept. CG
Altamonte Springs, FL 32716

Ceres Press
P.O. Box 87
Woodstock, NY 12498

EarthSave
706 Frederick St.
Santa Cruz, CA 95062-2205

Food & Water
Depot Hill Road
RR#1, Box 114
Marshafield, VT 05658

Pure Food Campaign
630-1130 17th St., NW
Washington, DC 20036

Vital Health
Burgundy Court Publishers
P.O. Box 8182
Fort Collins, CO 80526-8003

BIBLIOGRAPHY

Bacharach, Bert. *How To Do Almost Everything*. New York: Simon and Schuster, 1970.

Bines, Robin, and Peter Marchand. *Ecover Information Handbook; An Introduction to the Environmental Problems of Cleaning Products*. Mouse Lane, Steyning, West Sussex, BN4 3DG: Ecover, 1989.

Brown, Elizabeth A. "Toxics at Home - Hazardous-Waste House Cleaning." *The Christian Science Monitor*, January 3, 1990.

Buchman, Dian Dincin. *Herbal Medicine: The Natural Way to Get Well and Stay Well*. New York: David McKay Company, 1979.

Christensen, Karen. *Home Ecology: Making Your World a Better Place*. Golden, CO: Fulcrum Inc., August 1990.

Clean Water Fund. *Every Citizen's Environmental Handbook*. Washington, DC, 1990.

Dadd, Debra Lynn. *The Nontoxic Home: Protecting Yourself and Your Family from Everyday Toxics and Health Hazards*. Los Angeles: Jeremy P. Tarcher, 1986.

Dadd, Debra Lynn. *Nontoxic and Natural: A Guide for Consumers; How to Avoid Dangerous Everyday Products and Buy or Make Safe Ones*. Los Angeles: Jeremy P. Tarcher, 1984.

Environmental Hazards Management Institute. *Household Hazardous Waste Wheel*. Durham, NH, 1988.

Garland, Anne Witte. *For Our Kid's Sake: How to Protect Your Child Against Pesticides in Food*. New York, Mothers and Others for Pesticide Limits and the Natural Resources Defense Council, 1989.

Goldbeck, David, and Nikki Goldbeck. *The Goldbecks' Guide to Good Food*. New York: Plume/New American Library, 1987.

Golos, Natalie, et al. *Coping With Your Allergies*. New York: Simon and Schuster, 1979.

Gosselin, Robert, M.D., Ph.D., Roger P. Smith, Ph.D., Harold C. Holdge, Ph.D.,D. Sc., and Jeanne Braddock. *Clinical Toxicology of Commercial Products, 5th edition*. Baltimore and London: Williams and Wilkins, 1984.

Greenpeace, *Step Lightly on the Earth*. Washington, DC: Greenpeace, 1989.

Hampton, Aubrey. *Natural Organic Hair and Skin Care*. Tampa, FL: Organica Press, 1987.

Hassol, Susan, and Beth Richman. *Everyday Chemicals: Creating a Healthy World/101 Practical Tips for Home and Work*. Snowmass, CO: The Windstar Foundation, 1989.

Hiscox, Gardner D., ed. *Henley's Formulas for Home and Workshop*. New York: Crown/Avenel Books, 1979.

Hollender, Jeffrey. *How to Make the World a Better Place: A Guide to Doing Good*, New York: William Morrow and Co., 1990.

Hupping, Carol, Cheryl Winters Tetreau, and Roger B. Yepsen, Jr., with special assistance by Bobbie Wanamaker, ed. *Rodale's Book of Hints, Tips, and Everyday Wisdom*. Emmaus, PA: Rodale Press, 1985.

Hylton, William H., and Claire Kowalchik, eds.. *Rodale's Illustrated Encyclopedia of Herbs*. Emmaus, PA: Rodale Press, 1987.

Johnson, Jerry Mack. *Down Home Ways: Old Fangled Skills for Making Hundreds of Simple, Useful Things*. New York: Greenwich House, 1984.

Lassiter, William Lawrence. *Shaker Recipes and Formulas For Cooks and Homemakers*. Prineville, OR: Bonanza Books, 1959.

League of Women Voters of Albany County. *Household Hazards: A Guide to Detoxifying Your Home*. Delmar, NY: League of Women Voters of Albany County, 1988.

Libien, Lois, and Margaret Strong. *Super Economy House Cleaning*. New York: William Morrow and Co., 1976.

Lord, John. *Hazardous Wastes From Homes.* Santa Monica, CA: Enterprise for Education, Inc. 1988.

Mack, Norman, ed. *Back to Basics: How to Learn and Enjoy Traditional American Skills.* Pleasantville, NY: The Readers Digest Association, Inc., 1981.

McClelland, Maureen R. *Toxins in the Home and Safe Alternatives.* East Falmouth, MA: The New Alchemy Institute, 1989.

Moore, Alma Chesnut. *How to Clean Everything.* New York: Simon and Schuster, 1977.

New York State Department of Environmental Conservation. *Household Hazardous Waste Fact Sheets.* Albany, NY: Division of Hazardous Substances Regulation, 1988.

Periam, Jonathan. *The Home and Farm Manual.* New York: Greenwich House, 1984.

Rose, Jeanne. *Jeanne Rose's Herbal Body Book, The Herbal Way to Natural Beauty and Health for Men and Women.* New York: Grosset & Dunlap, 1976.

Stark, Norman. *The Formula Book.* Kansas City, KS: Sheed and Ward, 1975.

Stark, Norman. *The Formula Book II.* Kansas City, KS: Sheed and Ward, 1976.

Stuart, Penny. *"Back to Borax." Canadian Living,* November 1989.

U. S. Environmental Protection Agency, Office of Air and Radiation. *The Inside Story: A Guide to Indoor Air Quality.* Washington, DC: GPO, September 1988.

U.S. Department of Health and Human Services National Toxicology Program. *Fifth Annual Report on Carcinogens, Summary.* Washington, DC: GPO, 1989.

Wallace, Dan, ed. *The Natural Formula: Book for Home and Yard.* Emmaus, PA: Rodale Press, 1982.

ABOUT THE AUTHOR

Formerly a painter, Annie Berthold-Bond changed careers to become an environmentalist when a pesticide exposure left her chemically sensitive. She lives in Annandale, N.Y. with her husband, Daniel, and daughter, Lily.

Annie is the editor of *Green-keeping*, a newsletter that provides practical information about effective and accessible nontoxic and environmentally safe alternatives. To contact Annie, be included on her mailing list or if you have a nontoxic solution or question, write to:

Green-keeping
Box 110
Annandale, NY 12504